Risk, Choice and Prediction

AN INTRODUCTION TO EXPERIMENTATION

W. J. Youden

DUXBURY PRESS, North Scituate, Massachusetts

A Division of Wadsworth Publishing Company, Inc.
Belmont, California

ISBN-0-87872-055-3
L. C. Cat. Card No. -73-84604
Printed in the United States of America

1 2 3 4 5 6 7 8 9 10 - - - 78 77 76 75 74

Risk, Choice and Prediction was edited by Margaret Kearney.
The cover and the design were prepared by James Rigney.

To
D I D I

without whom this book might
never have come to pass.

Contents

Preface

Jack Youden was perhaps the best expositor statistics has ever known. In his widely acclaimed lectures and writings, he could take almost any aspect of statistics and explain it in a way that would leave his audience with some understanding of the subject *and*, more importantly, a desire to learn more about it. This is what he has done in his final work, *Risk, Choice and Prediction*. He has taken some ordinary everyday objects like a set of bingo cards, some marbles, and a few dice and woven a story describing how he went about discovering so much that would have been missed by the less-experienced eye. And, along the way in this highly readable story, he shows the reader how "some simple mathematics and a little helpful thinking" can provide a valuable asset to intuition when looking at experimental data. In so doing he reaches to the very heart of the statistical method.

Risk, Choice and Prediction is for anyone. Anyone, that is, who wants to learn in a relatively painless way how the concepts and techniques of probability and statistics can help us to better understand today's complex world.

At the request of Mrs. Youden, any royalties from the sale of this book will be donated to a scholarship fund for a deserving student at the Washington Technical Institute, which now occupies the buildings where Jack worked during his years at the National Bureau of Standards.

It has been my priviledge to work with Mrs. Youden, Thomas A. Willke, Mary G. Natrella, Richard J. DeGray, Margaret Kearney, and Alexander Kugushev, in seeing this book through the final stages of publication.

Brian L. Joiner
The Pennsylvania State University
11/1/73

Risk, Choice and Prediction

DATA
CONTINUE EXPERIMENT
POSSIBLE FAILURE
MORE HARD WORK
ABANDON EXPERIMENT
MAY MISS IMPORTANT DISCOVERY

Why do I write this book?

1

I write because my life as a scientist has convinced me of three things. First, that the ideas connected with words like ***chance*** and ***probability*** are extremely useful in physical science, in engineering, in agriculture, in biology, in medicine, in the social sciences and in economics. Second, I have observed that many workers in these fields have been handicapped by a lack of understanding of the laws of chance. And third, I believe that it is possible at an early age to begin to grasp these ideas.

The first half of my more than 40 years as a scientist was spent doing experiments. My training was in chemical engineering and physical science, especially in chemistry. A chemist knows how to make pure materials such as pure water, sugar, salt, copper, aluminum, aspirin and countless other substances used in households and industry. We, as users, take for granted that sugar is sugar, forgetting that the source may be sugar cane, sugar beets or maple tree sap. In every case the chemist knows how to refine the raw material and obtain the same sugar.

While the salt on the dinner table may come from the sea or from mines deep in the earth, if it is carefully purified we always get the same product. There is no variation in properties such as taste, color, odor or solubility between different samples of pure salt. This uniformity—the absence of variability—is a tremendous help to the chemist in his experiments.

Consequently, I was quite unprepared for the variability of living material when I joined a staff of research biologists. Quite possibly you have observed in gardens how the plants vary in size although all came from seeds apparently identical. This variation among plants of the same kind, grown under the same conditions adds to the difficulty in trying to show a difference in yields between two varieties of wheat, or beets or other crop. Such were the problems of these biological researchers who provided the initial stimulus to my study of variation and the laws of chance.

The same problems arise in engineering. Consider construction materials such as wood, steel, alloys, and cement, to name just a few. We might expect wood to vary as it comes from trees. Even with the greatest care different production batches of steel and alloys are not identical. So too, some allowance or provision must be made in building houses, airplanes and automobiles for the possible variation in strength of the materials used. Cement, a highly variable product used in building dams, buildings and bridges, is mostly made from mineral deposits which cannot be purified chemically. Therefore, the building of safe structures depends upon the use of mathematics to reveal the limits of variability of the cement being used.

In the field of medicine patients vary greatly in their reaction to drugs and other treatments. Healthy people also vary in their blood pressure, blood count, blood sugar, and many other measurements which doctors routinely make. Medical researchers are continually seeking improved treatments. The problem is to find whether or not a new treatment "N" is better than the currently used treatment "C". Again, the task is complicated by the variability of people. So medicine also makes great use of the mathematics of variation.

Most scientific work involves making careful, precise measurements. Therefore, a knowledge of the laws of variation is indispensable. These same ideas are involved in many commercial contracts. Payments depend upon the quality of the material received as determined by measurements made on samples from the material. The assessment of custom duties on wool and other products also encounters this same need to evaluate the variation in the material.

The second half of my scientific life has been as a consultant to other experimenters. This has given me the opportunity to work in a wide variety of fields. What impresses me now is that the laws of chance can be put to work in so many different specialities.* That is why the ideas in this book will be important to you if your life work is to be in science, either physical or social, in any field of engineering,

*A recent book that discusses a wide variety of applications is ***Statistics: A Guide to the Unknown*** by Tanur, Judith M., Mosteller, Frederick et al. (1972) Holden-Day, California.

in medicine, in business or in industry. Even areas like history, language and the arts are finding these concepts of increasing importance. Good research, today, requires a careful evaluation of the risk of being in error.

The laws of chance also affect our daily lives as human beings. We buy lottery tickets and place bets on horses and the outcome of sporting events. We buy fire, life, accident and health insurance and listen to weather predictions. We try hard to predict what may happen in this uncertain world, ordinarily without formally weighing the risk of making the wrong choice. Someday we may learn to be more careful about even these decisions.

One final word. As you read this book, please take time to do the experiments. In my opinion if you ***do*** the experiments I have suggested, you will learn more than by reading the whole book and doing no experiments. No one ever becomes a tennis player, a musician or an artist by reading books. You must get on the tennis court, practice your music, or use a paint brush to get anywhere. So please do the experiments, and even if your livelihood does not make use of risk, choice and prediction, your understanding of the world we all live in will hopefully be furthered by this book.

Randomization

2

Always, when the experimenter has obtained his data, he is faced with the problem of interpreting it. Often his results are clearly negative. The experimenter has not, as he hoped that he might, found a new test for the early detection of cancer. Occasionally, the evidence strongly supports success. A striking example was the large-scale test of the Salk polio vaccine conducted in 1959. Other times there is just a suggestion of positive results. The data are not convincing but they *may* be ground for a little hope. At this point the experimenter is faced with a tough choice. Shall he do some more of the same kind of work or shall he abandon his project and try something else? If he continues there is the possibility that this first hint was false and that the inevitable variability of animals and plants has tricked him. On the other hand, if he abandons this experiment, he may be turning away from an important contribution to science.

The experimenter needs the best possible guide to estimate the chance that the apparent hint of success could be a false alarm. Then he will know the risk he faces in making his choice. Not many people realize that prior to 1908 scientists were unable to calculate this risk. In that year an Irish chemist, W.S. Gosset, devised the first really useful guide while trying to improve the quality of certain crops. He used the pen name "Student," and some day in your education you will probably hear about the "Student's t test" that he developed. You might think all scientists would rush to use it, but not so. It was only by accident that I heard of it in 1925. During that same year 1000 copies of another book appeared, by Sir Ronald A. Fisher.* Despite its being a difficult book to read, in the years

*R.A. Fisher, ***Statistical Methods for Research Workers***, Oliver and Boyd, Edinburgh, 1925.

following its appearance scientists of all kinds went to study with the author. (I was fortunate enough to be one of them.) Since then, many hundreds of books have been written to explain and supplement Fisher's historic volume. Several societies were formed, journals were established, and new departments were added to universities. Yet even today few students are introduced to the subject of statistics until they are far along in their studies.

I wrote this book because I am convinced that this way of thinking should be a part of the early training of all who think they might do experimental work or use the experimental work of others. It might be research work, medicine, electronics, computers or the environmental sciences, as well as all the older disciplines. A scientist does not spend his time blindly "trying things to see what will happen." Laboratory space and equipment are too expensive to be used so inefficiently. Instead the scientist makes use of prior work (his own and others) to make a very tentative picture of the situation.

Perhaps the medical researcher wishes to make a new and improved drug. First he will surely examine the ones presently available. After thinking of some modification that is worth making, he will find out how similar compounds are made and eventually produce the new compound, "N." Then comes the crucial problem of testing it in comparison with the best or the most commonly used drug, which is often called the control, "C."

One measure of the effectiveness of a drug might be the length of stay in a hospital. Of course, people differ from one another and in the severity of their afflictions. Consequently, the length of stay varies even when the same treatment is given. This variation tends to obscure any reduction in length of stay in the hospital which might be due to the new drug. Obviously, the researcher will need a number of patients, some fair method of assigning them to the two drugs, and some method to help him interpret the data.

Suppose that for a preliminary experiment, we decide to use eight patients: four to receive drug N and four to get the control drug C. A sticky question immediately arises. Who is going to assign the patients to the drugs? If we do the job, then we are wide open to challenge by the researcher who developed the currently-used drug. If we report that our drug is better, the other researcher might claim that we selected the most ill patients for his old drug. We deny this. But wait a minute. Would we be willing to let ***him*** assign the patients? Of course not. Now you can see that we need some impartial method of assignment. We must guard against either conscious or unconscious bias. So at last we come to the subject of this chapter.

The standard, and indispensable step at this point is to make a random assignment. Even then some care must be taken. We should avoid knowing the order lest we be open to the charge that we were influenced in the time we sent our patients to the hospital or in the time when we sent them home. In some way a random order of four N's and four C's must be obtained by someone who is not involved. Eight envelopes are prepared and numbered one to eight. The letters N and C, in their random order, are sealed in the eight envelopes. Then, as each patient arrives, an envelope is opened and the letter inside determines the drug to be given. Randomization avoids personal bias. Furthermore, randomization makes possible the assessment of the risk of claiming that there is a difference between the drugs when actually there is no difference. Since patients' attitude can affect the length of their hospitalization, it is important that they not know which drug they are receiving. In fact, in most experiments, the doctors are not even told which patients are getting which medicine.

No doubt we would feel elated if all four patients taking our drug, N, spent less time in the hospital than any of the control patients. That is, when we rank the patients in order, from shortest stay to longest, we would like to have the result NNNNCCCC. Have you any idea of the risk of getting this order if, in fact, both drugs are equally effective? We can find out about this risk in two ways, either by a bit of mathematics or by doing some experiments with marbles. We will do both.

Mathematically, the number of different orderings of the eight letters is

$$\frac{8 \times 7 \times 6 \times 5 \times 4 \times 3 \times 2 \times 1}{(4 \times 3 \times 2 \times 1) \times (4 \times 3 \times 2 \times 1} = 70.$$

A good exercise is to enumerate all 70 possibilities. (For example, NNNNCCCC, NCNCCCNN, CCCNCNNN, and so on.) You will need some logical ordering or you will probably miss some and duplicate others.* Really there are only 35 you need to find because you can get the other 35 by interchanging the N's and the C's. (Did you

*See Appendix Table A for complete enumeration.

discover that only two of these 70 orders show a complete separation of the letters?) What we are leading up to, is that if there is no difference between the drugs, then the random assignment of the letters to the patients means that there has been the same opportunity for any of the 70 orders to be found.

Another way to investigate random order is by experimenting with marbles. To make the simple device shown below, two small strips of wood are nailed together to make a trough. One end is blocked by a small piece of wood. The other end is supported by a small block of wood to give the trough a slight incline. I bought four marbles with red centers and four with yellow centers, all the same size. I shook the eight marbles together in a plastic cup and poured them into the trough to form a line. (I had previously marked the eight positions from one to eight.) I was now ready to do the experiment.

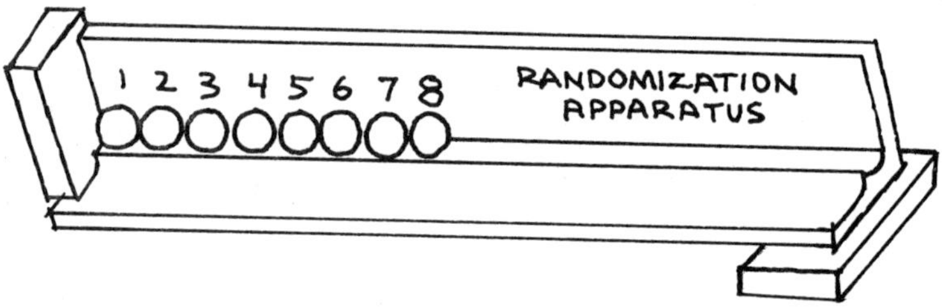

My data record is opposite. As you see, I made a chart with columns marked one through eight. All I had to do was read the numbers for the red marbles and place four dots under the same numbers on my data sheet. The first trial showed the red marbles in positions 3, 4, 6 and 7, so the dots were entered accordingly. Each row corresponds to a trial, and they are listed in three sections. I did 70 trials in all.

At this point a question might be raised about the apparatus. Should we have depended on this apparatus to do a good job of providing experiments with a random order in which to assign the drugs? Using a little arithmetic we can easily check the performance of this particular apparatus. There are four red marbles, and we made 70 trials. That makes a total of 280 appearances of red marbles. Surely you will agree that these inert marbles, *if* they are alike, should end up just as often in all eight positions in the trough. That means we predict red marbles to turn up 35 times in each position –not exactly 35 but not too far from 35 either. By counting the dots on my record sheet, I obtained the counts shown at the top of p. 10.

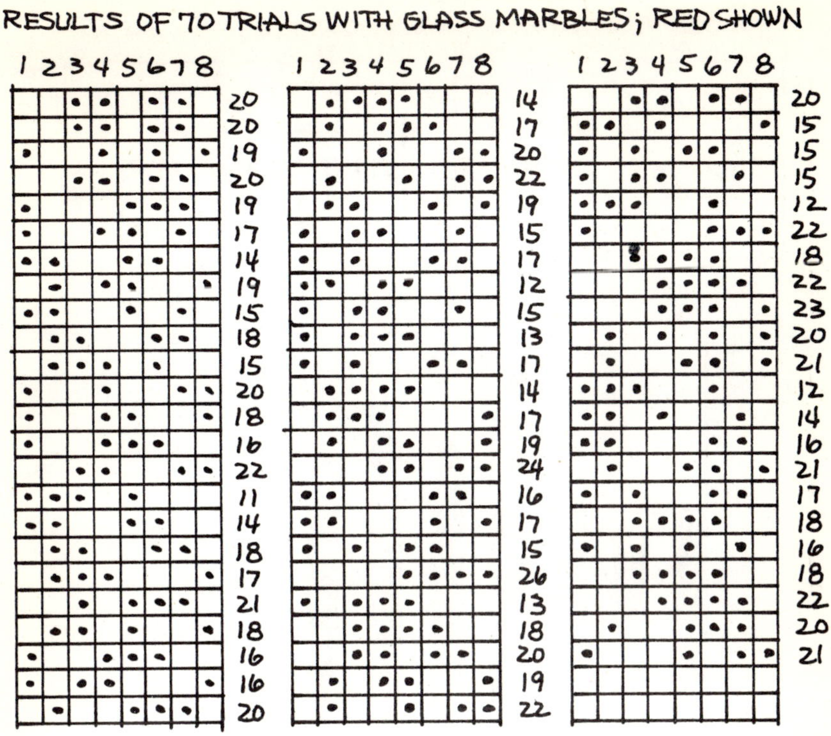

They are also shown graphically on p. 10. These turned out to be acceptably close to prediction.

We can make two performance tests. The data record above shows the sum of the position numbers for each trial. For example, the first trial had a sum of 3 + 4 + 6 + 7 = 20. In real research this number is often used to evaluate the data. The sum of the four numbers (see below) cannot be less than 10 (red marbles in first

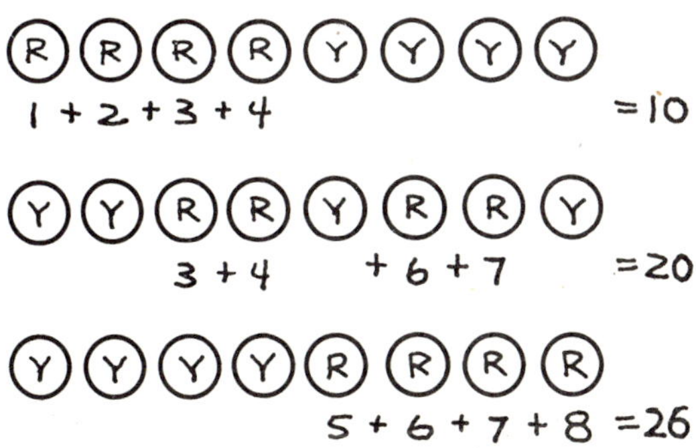

TROUGH POSITION AND TIMES OCCUPIED; RED GLASS

Position	1	2	3	4	5	6	7	8	Total
Observed	34	32	35	40	39	41	34	25	280
Predicted	35	35	35	35	35	35	35	35	280

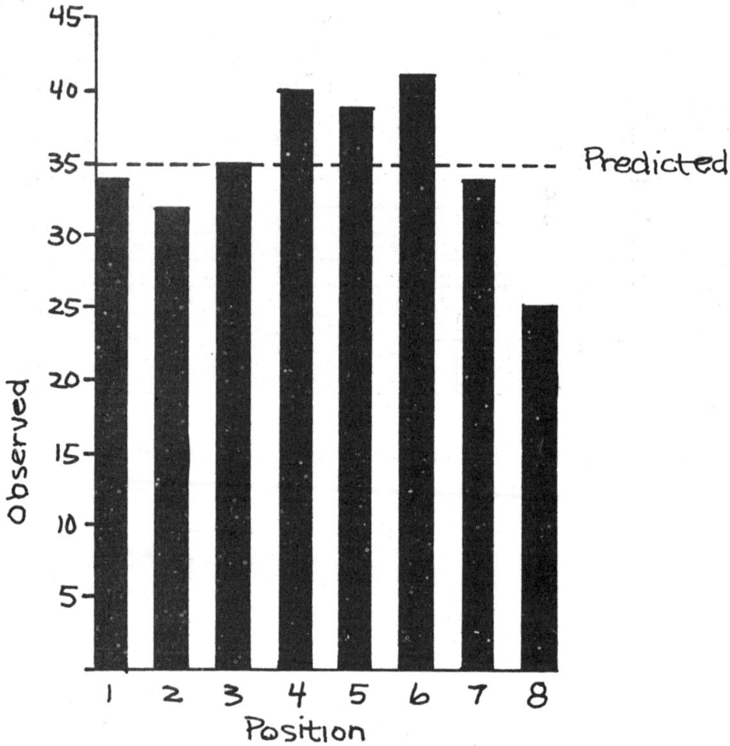

TROUGH POSITION AND TIMES OCCUPIED; RED GLASS

four positions) or more than 26 (the sum of the last positions: 5 + 6 + 7 + 8).

The customary way to tally such a collection of scores is shown opposite. All possible scores are listed in a column. Then go down the data record and make a stroke opposite the proper score for every sum found in the data record. The fifth stroke is usually made hori-

DISTRIBUTION OF SCORES;
RED GLASS MARBLES

Score	Observed	Predicted
10		|
11	|	|
12	|||	||
13	||	|||
14	𝍸	𝍸
15	𝍸 |||	𝍸
16	𝍸 |	𝍸 ||
17	𝍸 |||	𝍸 ||
18	𝍸 |||	𝍸 |||
19	𝍸 |	𝍸 ||
20	𝍸 𝍸	𝍸 ||
21	||||	𝍸
22	𝍸 |	𝍸
23	|	|||
24	|	||
25		|
26	|	|
Total	70	70

zontally to tie the first four strokes together in a compact bundle. In the second column, just to the right of the scores for the glass marbles, I entered the predicted frequency for each of these possible scores. Visual comparison shows the red glass experiment agrees rather well with prediction.

If you enumerated the 70 possibilities as I suggested for the drug experiment, you can quickly get the predicted frequencies. First you add up the position numbers for the N's and then tally as in the table above. The only 10 should come from the combination NNNNCCCC. The two ways to get a 12 are NNNCCNCC and NNCNNCCC.

Because the red and yellow marbles did not contrast very well in a photograph I looked around for some others. I found some blue and green clay marbles which were used as markers in the game of Chinese checkers. I decided to repeat the experiment using these

RESULTS OF 70 TRIALS WITH CLAY MARBLES; BLUE SHOWN

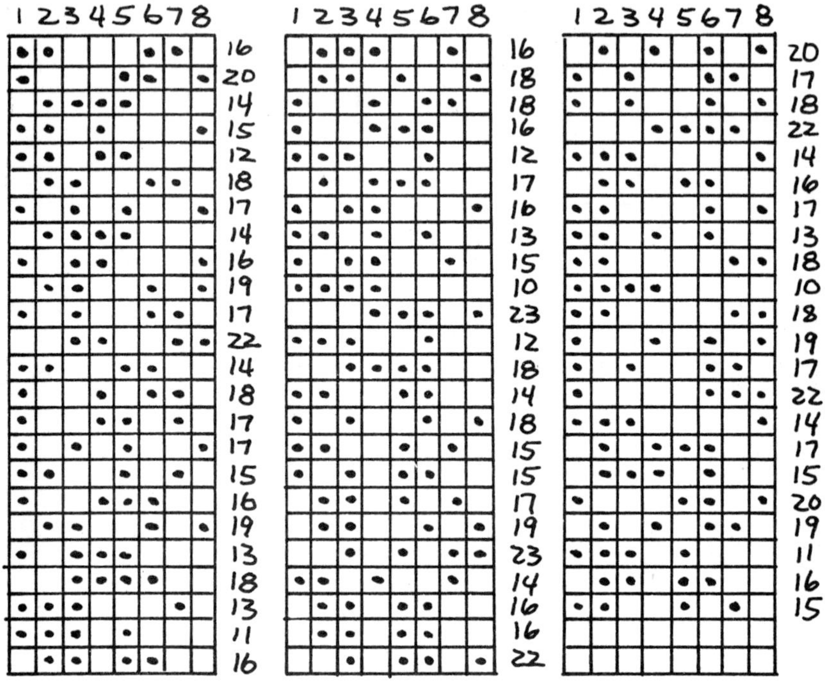

blue and green clay marbles. I recorded the positions of the blue clay marbles above. I checked the results by counting up the times the blue marbles appeared in the eight positions and obtained the following results:

TROUGH POSITION AND TIMES OCCUPIED; BLUE CLAY

Position	1	2	3	4	5	6	7	8	Total
Observed	44	42	41	30	34	41	23	25	280
Predicted	35	35	35	35	35	35	35	35	280

These results are plotted on p. 13. The distribution was more uneven for the clay marbles. When I tallied the results at the top of p. 14, the distribution of the scores was also quite disturbing.

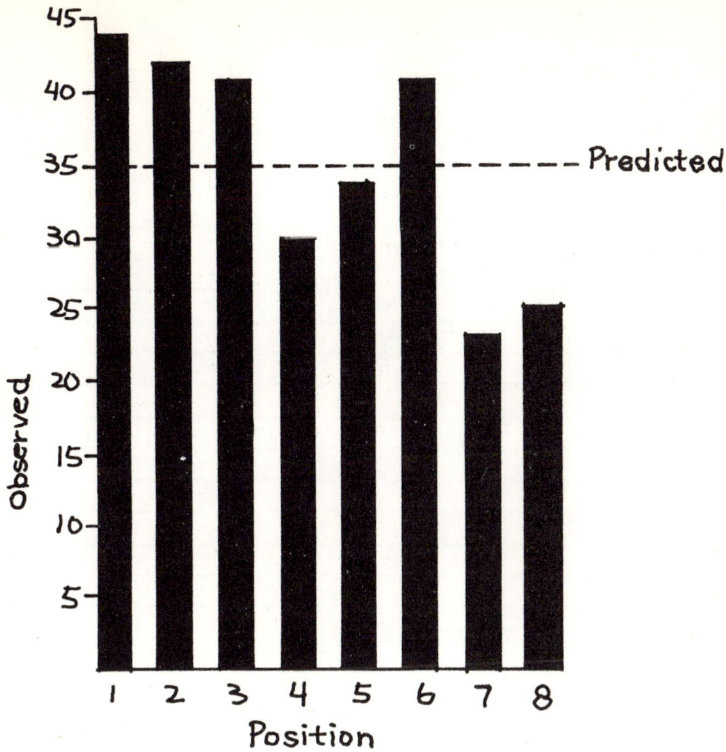

TROUGH POSITION AND TIMES OCCUPIED; BLUE CLAY

There is a mathematical test that is easy to apply to the position data shown on p. 10 and p. 12. As you can see on p. 15, the differences from prediction (for example, 35 - 34 = 1) are first squared, then divided by the predicted values ($1^2 \div 35 = 1/35$). These fractions are then added separately for each type of marble to give what is called their *Chi square values*. The glass marbles gave a sum of 188/35 = 5.4, and the sum for the clay marbles was 472/35 = 13.4. Close agreement between the observed and predicted values will give a small value of Chi square, while substantial disagreement between the observed and predicted values will be indicated by a large Chi square value. Mathematical tables give 1.7 to 16.0 as a reasonable range in which we may expect Chi square to vary in the situation we have here when there is no real difference between the colors. Since 5.4 and

DISTRIBUTION OF SCORES;
BLUE CLAY MARBLES

Score	Observed	Predicted
10	||	|
11	||	|
12	|||	||
13	||||	|||
14	~~||||~~ ||	~~||||~~
15	~~||||~~ ||	~~||||~~
16	~~||||~~ ~~||||~~ |	~~||||~~ ||
17	~~||||~~ ~~||||~~	~~||||~~ ||
18	~~||||~~ ~~||||~~	~~||||~~ |||
19	~~||||~~	~~||||~~ ||
20	|||	~~||||~~ ||
21		~~||||~~
22	||||	~~||||~~
23	||	|||
24		||
25		|
26		|
Total	70	70

13.4 both fall inside this range, we have no significant evidence of a difference between colors.*

I did try another way of looking at the data. First, I took the red marbles. Consider the first four positions—1, 2, 3, and 4. In any given trial there might not be any red marbles in these positions. Still, there might be one, two, three, or even four marbles. So I went over the 70 trials again and counted up how often each of these five possible events occurred. The results are shown on p. 16. Now we need a set of prediction values for comparison. Imagine that all four red glass marbles occupy positions 1, 2, 3, and 4. There is only one way for this to happen. Next, suppose we figure out how many ways we could arrange to have three of the positions occupied. We could remove any one red marble from the positions 1, 2, 3, and 4, and it could substitute for a yellow marble in any one of the four positions 5, 6, 7 and 8. This action leads to a total of 4 X 4 or 16 ways to obtain trials with a red marble in some three of the first four

*A more complete description of the Chi square test is given in most elementary statistics books.

GLASS MARBLES

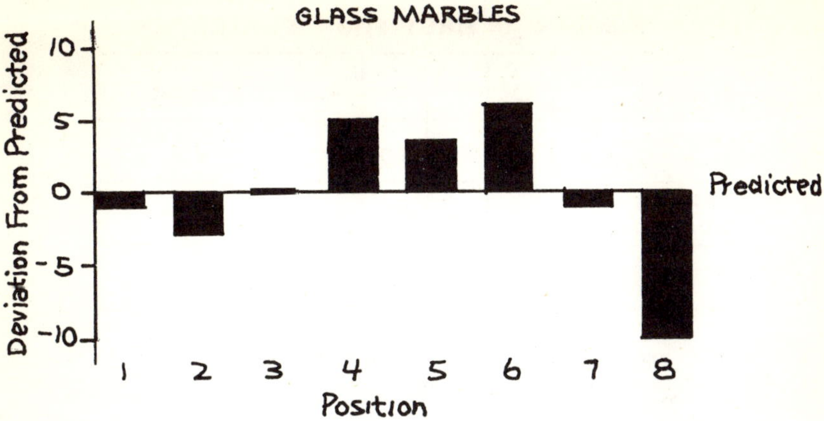

$$\frac{-1^2}{35}+\frac{-3^2}{35}+\frac{0^2}{35}+\frac{5^2}{35}+\frac{4^2}{35}+\frac{6^2}{35}+\frac{-1^2}{35}+\frac{-10^2}{35}=\frac{188}{35}=5.4$$

Chi Square Value

CLAY MARBLES

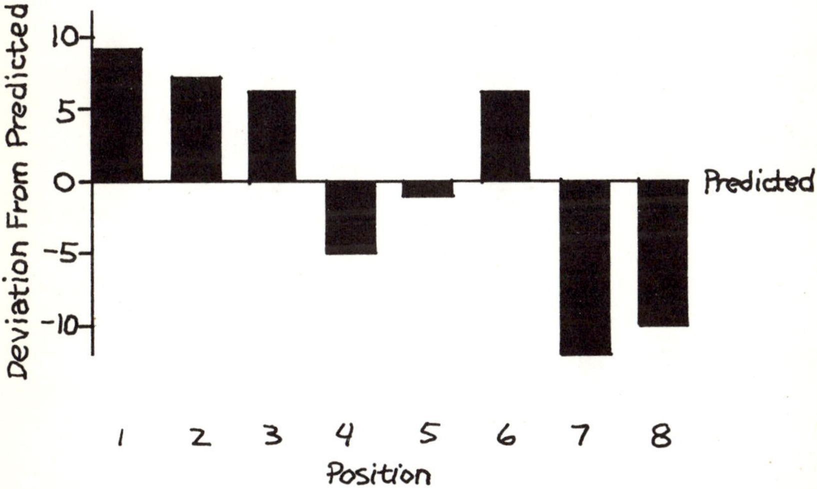

$$\frac{9^2}{35}+\frac{7^2}{35}+\frac{6^2}{35}+\frac{-5^2}{35}+\frac{-1^2}{35}+\frac{6^2}{35}+\frac{-12^2}{35}+\frac{-10^2}{35}=\frac{472}{35}=13.4$$

Chi Square Value

NUMBER IN FIRST FOUR PLACES

Number	0	1	2	3	4	Total
OBSERVED RED GLASS	1	12	40	17	0	70
OBSERVED BLUE CLAY	0	8	40	20	2	70
PREDICTED	1	16	36	16	1	70

positions. To leave two red in the first four positions we must remove two. There are just six ways to pick that pair from the four, and it is possible to put this pair into the positions 5, 6, 7 and 8 in six different ways. Altogether then there are 6 X 6 or 36 different ways to have trials with just two red in the first four positions. Why don't you check the other two possibilities, that is to leave one red, and then no reds.*

The red glass marbles show a much better agreement with prediction than do the blue clay. I decided I should repeat all the work, and I summarized the findings in the tables on p. 17, under the headings "Experiment 2." (The results of the earlier experiments are repeated next to the headings "Experiment 1.") The glass marbles again conform better to prediction. Then the two experiments are combined (next to the heading "Experiments 1 and 2 combined"). The contrast between the two kinds of marbles is even more evident now. The frequency with which the different positions are occupied again shows the clay marbles departing from prediction in about the same way as in the first experiment. We now have twice as much information as we had before. We predict 70 in each position. When a Chi square test is made on the combined data from the two experiments the value for the glass marbles lies well inside the limits given above. The clay marbles give a Chi square of 24.2. (You should check it.) This result is quite unacceptably high, giving us significant evidence that the blue marbles differ from the green ones.

Early workers discovered that it was no easy matter to devise a satisfactory randomizing device. After much effort, tables of random numbers were constructed. Today experimenters use such tables rather than construct their own. It had seemed to me that these

*See Appendix Table B for an enumeration of the other possibilities.

TROUGH POSITION AND TIMES OCCUPIED

	Position	1	2	3	4	5	6	7	8	Predicted
Experiment 1	OBSERVED RED GLASS	34	32	35	40	39	41	34	25	35
	OBSERVED BLUE CLAY	44	42	41	30	34	41	23	25	35
Experiment 2	OBSERVED RED GLASS	39	36	41	38	36	28	29	33	35
	OBSERVED BLUE CLAY	46	36	33	49	33	36	30	17	35
Experiments 1 AND 2 combined	OBSERVED RED GLASS	73	68	76	78	75	69	63	58	70
	OBSERVED BLUE CLAY	90	78	74	79	67	77	53	42	70

NUMBER IN FIRST FOUR PLACES

	Number	0	1	2	3	4	Total
Experiment 1	OBSERVED RED GLASS	1	12	40	17	0	70
	OBSERVED BLUE CLAY	0	8	40	20	2	70
Experiment 2	OBSERVED RED GLASS	0	7	44	17	2	70
	OBSERVED BLUE CLAY	0	4	40	24	2	70
	PREDICTED	1	16	36	16	1	70

eight marbles should be random enough for this small experiment. The fact that the glass marbles did check the predictions seems to have cleared the device and the method of using it. The clay marbles must be the culprit.

I lacked any precise means of measuring weight or diameter of the clay marbles. I did line up all eight in a row and placed a straight edge on top of them. I could see enough daylight above the green ones to establish that the blue ones were larger. So a plausible explanation for the excess of blue marbles in the first part of the line-up is that the larger blue ones "elbowed" their green cousins out of the way. A detailed investigation of this diameter effect could be undertaken. For our purposes, however, it is enough to see how easily an apparently good randomizing procedure can go wrong. We owe the clay marbles a vote of thanks for reminding us to be on our guard.

I have said very little about the scores obtained in each trial. You probably noticed that the minimum score of ten resulted when the first four places were occupied by the same color. Even a score of ten is not so very exceptional since it could happen by chance one time in 70. Larger numbers of trials are needed. Likewise, the evidence for a difference between a new drug and a control could be improved by using more patients. For example, if we used 16 patients, eight for each drug, the minimum possible score is the sum of the first eight positions, or 36. It is unlikely that such a low score will be achieved if the two drugs are identical. With 16 good marbles the probabilities indicate that we should expect the first eight positions to be occupied by the blue marbles only once in 12,870 trials. If in one trial with 16 patients all the new drug patients did better than the control patients, it could be that one in 12,870 chance. However, most will agree that the evidence is strong enough to cause us to abandon our tentative assumption that the two drugs have the same potency. We would feel quite safe in concluding that the new drug really is the better one.

Scientists generally are not as exacting. A score as low as 44 or less would be shown by the 16 marble test only once in 100 trials. Such an exceptional event would surely lead to further research. What is needed, in order to save experimenters a lot of arithmetic, is a table listing the scores that are low enough to arouse the interest of experimenters. Happily such a scoring technique was invented by a chemist colleague who worked in the laboratory next to mine at the biology research institute. His name will not be forgotten because his scoring scheme is known as the Wilcoxon rank sum test. His technique is widely used in judging the results of experiments in medicine and other important areas.

One day, a friend of mine, Dr. Dick DeGray pointed out to me an interesting feature of my marble data. His note says "In both experiments, with both clay and glass marbles, the sum of positions 7 and 8 is the lowest of all pairs of positions. I know the Chi square test shows nothing significant, even when the pairs of positions are tested, but the 7, 8 pair is low four times out of four. Why?" This is a proper question, but I cannot think of any reason for the outcome. If I still had the marbles, more experiments could be made to see if low values for these two positions persisted. I point out that these two positions gave high counts for the green clay marbles and the yellow glass marbles. Quite possibly I picked colors to record that had a tendency to be first in line for both kinds of marbles. Had I elected to record the positions of yellow glass marbles, rather than the red ones, this question would never have been raised. In that case the table on p. 17 would have shown the 7, 8 pair to be high on the glass marble experiments and low for the clay marbles. So I do not think the question raises any serious worries about the apparatus or technique. The marbles may well have lacked uniformity, and I was just unlucky in my choice of colors.

Nevertheless the question led me to a further experiment. For many years I have used for demonstrations the apparatus that is sketched below:

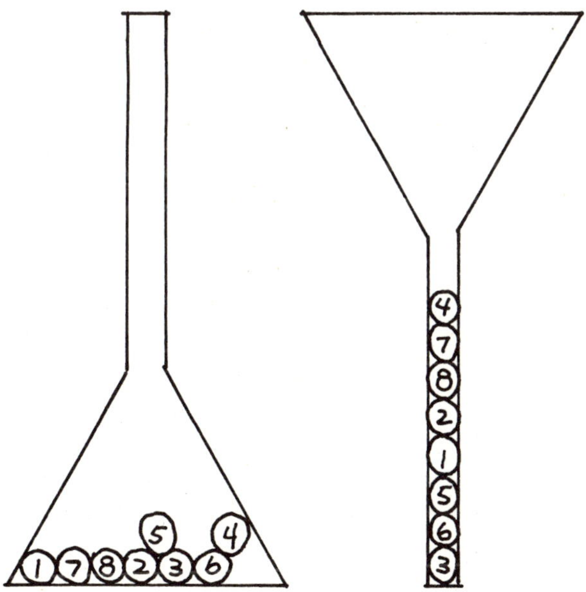

It consists of a small Erlenmeyer flask that has a length of glass tubing sealed to the neck. For the first model I used a short piece of rubber tubing to join the flask and tube. Usually I had nine green and nine red wooden beads in the flask. The beads are just a little smaller than the diameter of the elongated neck. The apparatus had the merit that it was easy to shake up the beads, then invert the flask while still agitating the beads, until all were lined up in the neck. The device was easy to use and gave a visual demonstration of the randomizing process. In my demonstrations many people were surprised at the frequency with which four or more beads of the same color were found together.

For the next experiment, I used this device with eight of the green beads now numbered from one to eight. Instead of just two colors, which did not make it possible to identify individual objects, I could now keep track of the performance of the eight individual beads. If the observed high (or low) counts in the marble experiment could have been connected with particular individual marbles, we might have been able to find out why or how they differed from their fellows.

I made 80 trials and recorded the bead numbers for each of the eight positions along the neck. The positions were numbered in reverse order so that position eight was occupied by the first bead out of the flask. We can examine the 80 trials and find out if a particular bead favored one position or even a small group of neighboring positions. The data are summarized opposite. The eight columns refer to the eight positions. The eight rows give the eight numbers of the beads. Each cell contains two entries. The first entry, 14 for bead one in position one, refers to the number of times that bead landed in position one. The second entry, 12, is the number found in a repeat experiment also of 80 trials. The lowest combined score of 11 is found for bead five in position four. The largest combined score, 33, is made by bead seven in position seven. Lucky seven?

The average of all the entries for each experiment must be ten. There are eight beads in every one of the 80 trials for a total of 640 bead appearances. As there are 64 cells, the average for all the cells must be ten. We shall be concerned, then, if we observe what appears to be a pattern among the entries for Experiment 1, and then find approximately the same pattern among the cell entries for Experiment 2.

RESULTS OF TWO EXPERIMENTS WITH THE FLASK

Bead Number	Position Number 1		2		3		4		5		6		7		8	
1	14	12	8	9	11	9	4	11	17	7	7	9	11	13	8	10
2	7	9	4	14	14	10	18	13	12	11	9	9	9	8	7	6
3	11	10	11	10	15	9	5	12	6	11	16	12	9	7	7	9
4	9	10	9	13	7	6	18	9	8	12	11	12	7	8	11	10
5	13	11	7	12	9	12	6	5	12	8	16	8	7	12	10	12
6	8	10	13	7	8	12	12	11	9	10	5	7	17	8	8	15
7	9	6	17	8	6	15	8	6	5	8	10	10	14	19	11	8
8	9	12	11	7	10	7	9	13	11	13	6	13	6	5	18	10

Remember that the predicted entries are ten for each cell. The actual cell values are either larger than ten, equal to ten or smaller than ten. Examine the two values in a cell. Skip for the moment cells where an entry is exactly ten. Now either both entries exceed ten, both are less than ten, or one entry is more and the other less than ten. While it is not quite true, we will assume that an entry is as likely to be more than ten as it is apt to be less than ten. That leads us to think of it like tossing two coins. Either they come down both heads, both tails, or one head and one tail. The chance of their being alike is the same as their being unlike, that is 50-50. Going over all 64 cells we find 26 where the entries depart from ten in the same direction and 26 where they depart in opposite directions. So we have no basis for suggesting that an entry for the first experiment is any guide as to what we might anticipate the second experiment will show. Twelve cells contain at least one entry of ten. In six of these the other cell entry is greater than ten; in five the other entry is less than ten. One cell has both entries equal to ten. So these 12 cells certainly show no tendency to give any predictive clue given one entry. The beads all seem to line up in a sufficiently random order.

All the above is not nonsense. In an authoritative and widely used book on experimental design, the final chapter deals exactly with this problem. The authors, Cochran and Cox,* tabulated 1000 random arrangements of the numbers one through 16. Although their method was based on tables of random numbers, the authors felt it necessary to make sure that their orders were acceptably random. Experience has shown that it is far from easy to devise an effective randomization procedure.

Perhaps you can think of a randomization device you would like to try. Then do an experiment to see how random it really is.

*William G. Cochran and Gertrude M. Cox, ***Experimental Designs***, John Wiley & Sons, New York, 1950.

Lotteries
3

A lottery is defined as "a scheme for the distribution of prizes by lot; especially such a scheme in which lots, or chances are sold." People speak of drawing lots as a fair method of dividing up various objects. Drawing lots also serves to fix the order of a group, such as the position of horses at the starting gate. Drawing lots is generally regarded as a fair method that avoids favoring any of the participants.

The United States had an example in 1969 of a badly bungled lottery. The draft lottery drew a lot of criticism from statisticians, and justly deserved it. The whole affair reflected the naiveté about these matters that exists with many of my generation. Unfortunately, these senior officials learned the hard way that a large-scale lottery presents difficulties, and in 1970, stung by the criticism, they sought professional help.*

Many situations can be described as lotteries. For instance, there must be something worth winning (or avoiding), and nearly always it costs something to buy a chance. Door prizes, the cost of which is concealed in the price of admission, are common. Charities sometimes use a lottery as a means of raising money. In my younger days lotteries were characterized as a form of gambling, which really they are. They are considered illegal in some places, although today some states run them as a means of raising money. (You might try to find out what percentage the various state lotteries pay out in prizes and make in profits.) The British government uses a lottery to encourage purchase of its bonds. One widely operated lottery is the "numbers racket." It is illegal chiefly, I suspect, because the authorities do not get a cut of the profits. Another popular lottery is the game of bingo.

*An interesting pair of articles on the draft lottery appeared in *Science* magazine, 22 January 1971, volume 171, pages 255–261 and 306–308.

A lottery must not only be fair, but it also must not involve any skill on the part of the participants. Betting at the race track and participation in the immensely popular football pools are not really lotteries. They provide an opportunity for the skilled to make better use of information about the past performance of horses or teams.

Let's demonstrate a simple lottery. We need a bowl containing 20 white beans. Mark one with a red spot, one with a blue spot, and three with green spots. Get 20 of your friends to each give you a penny. Then have each one draw, without looking, one bean. The red bean pays ten cents, the blue bean, four cents, and the green beans, two cents each. Just enough money has been collected to pay the winners. Five participants have improved their resources and 15 are poorer.

To make the game a little more realistic, reassemble the 20 beans and add a further 30 unmarked beans. If these are all sold for a penny each, the lottery will take in fifty cents and pay out only twenty cents. The amount of profit can be controlled and in many instances is very large. The Irish Sweepstakes has financed some splendid hospitals. Notice, too, that if all the tickets are not sold, the larger prizes may not be drawn at all. Some schemes have drawn criticism because the largest prizes were never won!

One thing is certain. The temptation to risk a small amount for a possible large gain knows no boundaries. All nations, all races have many members who indulge in this activity. The fact that the sum of all gains is much less than the sum of all losses, either is not understood or at least is no deterrent. While it makes sense that the player should be sure he can afford to lose, most players can accept small losses, especially knowing a win makes possible an otherwise unattainable goal. Many people consider various forms of gambling as recreation, and I can see no serious objection to that point of view. Of course, like anything else the activity can go beyond good sense and bring the addict to disaster.

We have said that the game of bingo is a lottery because no skill is required. No doubt one reason for its appeal is that the outcome is quickly decided, and repeat games come in rapid succession. I have never played the game, although I had somehow acquired a fair understanding of it (probably at an army post). Evidently small versions are sold as games, and I found a set while vacationing in

Norway. In this game each person gets a card with 24 numbers on it. The center is marked free which means that it can be covered immediately by the player. The numbers run from one to 75. The first column contains a selection of five numbers from the first 15 numbers and so on for each column. This restriction makes it easy to locate a number if it is on your card. The master of ceremonies picks numbers at random from an initially complete set. The object of the game is to get five squares in a horizontal row, in a vertical row, or in either diagonal. As a number is called out each player places a wooden marker on the number if he has it. Chance determines the numbers called so no skill is needed.

There were 20 cards in this particular set. They are shown on p. 26. I wondered if some system existed for determining assignment of the numbers to the 24 squares. My quick opinion was that it would be rather difficult to determine the answer to a question such as "How many numbers are usually needed to achieve a winning card?" Of course, the number of cards in the game is important. The more cards, the sooner a winning combination will be found.

How were the numbers assigned to the columns? The first column (and every column except column three) is a set of five numbers drawn from 15 possible numbers. It is interesting to find out how many different sets of five can be formed from 15. The formula is:

$$\frac{15 \times 14 \times 13 \times 12 \times 11}{1 \times 2 \times 3 \times 4 \times 5} = 3003$$

I think you can figure out the method. Check it on some simple case where you can easily write out all the possibilities. For example, you can pick two from four objects in $(4 \times 3)/(1 \times 2) = 6$ different ways and three from five in $(5 \times 4 \times 3)/(1 \times 2 \times 3) = 10$ ways. Try it! This formula is very useful, and we will meet it again.*

I decided to count how often each number (from one to 15) appeared in the B column for all 20 cards. There will be a total of 100 counts. A convenient way to do this is shown in the first column of the table on p. 27. To my surprise there was no 5 on any card. There was only one 3. There were thirteen 11's. My years of experience gave me the feeling that this was too uneven an assignment of the 15 numbers to be the result of chance. Of course I did a little mathematics to back up this feeling. The calculations confirmed my

*Some results are given in Table B of the appendix.

COMPLETE SET OF BINGO CARDS

B	I	N	G	O
9	30	32	57	75
10	28	34	60	67
2	25	•	47	72
15	22	39	52	61
1	27	33	49	64

B	I	N	G	O
11	28	42	52	68
14	21	43	53	67
6	27	•	56	74
15	23	41	57	73
7	22	34	59	71

B	I	N	G	O
2	28	43	49	73
13	24	36	57	61
6	18	•	60	69
11	22	35	53	65
4	30	42	58	71

B	I	N	G	O
4	24	39	46	66
14	27	42	51	71
2	18	•	49	67
6	23	38	48	63
12	25	31	54	61

B	I	N	G	O
6	22	34	53	72
11	27	38	56	71
4	28	•	52	69
13	21	42	58	67
14	24	32	59	68

B	I	N	G	O
14	20	32	52	71
10	27	42	55	64
7	23	•	59	69
11	28	34	56	72
15	25	33	53	67

B	I	N	G	O
1	22	36	53	62
14	27	32	49	65
3	28	•	48	68
7	29	33	60	74
9	25	40	46	72

B	I	N	G	O
12	20	36	46	75
15	26	32	53	64
1	27	•	55	66
14	24	37	54	68
13	28	29	58	65

B	I	N	G	O
11	26	34	60	70
9	23	41	52	65
13	25	•	47	62
7	22	40	57	75
4	20	33	51	73

B	I	N	G	O
6	26	32	49	69
11	22	38	47	72
4	23	•	59	61
15	18	34	58	64
8	20	42	55	75

B	I	N	G	O
8	22	32	51	75
7	25	31	49	67
13	23	•	50	73
6	26	34	58	63
4	27	37	48	66

B	I	N	G	O
4	22	39	57	62
11	23	40	55	72
7	27	•	59	67
6	18	41	51	71
9	25	31	56	68

B	I	N	G	O
15	23	35	50	63
11	25	43	52	65
1	26	•	53	74
14	21	33	56	72
7	22	38	48	71

B	I	N	G	O
11	23	39	59	65
15	26	36	56	70
8	25	•	50	74
10	28	42	48	73
7	30	41	54	62

B	I	N	G	O
10	20	33	50	72
1	24	41	59	74
12	27	•	58	63
13	26	32	46	75
15	25	34	53	62

B	I	N	G	O
8	21	38	46	72
14	25	35	57	62
10	22	•	59	74
9	27	43	60	67
15	18	39	58	70

B	I	N	G	O
7	25	37	58	72
11	30	38	59	69
10	23	•	55	68
9	24	43	56	64
13	27	35	60	67

B	I	N	G	O
10	21	41	57	71
1	28	40	60	67
12	27	•	56	62
6	20	33	53	65
11	25	42	50	72

B	I	N	G	O
7	28	36	57	72
14	21	40	54	64
13	20	•	59	68
15	27	38	60	65
11	23	42	56	71

B	I	N	G	O
15	27	34	58	73
13	20	42	56	69
1	22	•	59	67
11	28	36	52	64
14	24	40	54	68

guess that to have a number totally missing was unusual if chance alone operated. About once in 60 times it could happen according to mathematics. And only about once in a 100 columns would there be such an uneven assignment. I won't spell out the mathematics here. The important thing for you is that there is an arithmetic that

COUNTS FOR EACH OF THE 75 NUMBERS ON BINGO CARDS

B	I	N	G	O
1 𝍸 ||	16	31 |||	46 𝍸	61 ||||
2 |||	17	32 𝍸 |||	47 |||	62 𝍸 ||
3 |	18 𝍸	33 𝍸 ||	48 𝍸	63 ||||
4 𝍸 ||	19	34 𝍸 𝍸	49 𝍸 |	64 𝍸 ||
5	20 𝍸 |||	35 ||||	50 𝍸	65 𝍸 |||
6 𝍸 |||	21 𝍸 |	36 𝍸 |	51 ||||	66 |||
7 𝍸 𝍸	22 𝍸 𝍸 |	37 |||	52 𝍸 ||	67 𝍸 𝍸 |
8 ||||	23 𝍸 𝍸 |	38 𝍸 ||	53 𝍸 ||||	68 𝍸 |||
9 𝍸 |	24 𝍸 ||	39 𝍸	54 𝍸	69 𝍸 |
10 𝍸 ||	25 𝍸 𝍸 |||	40 𝍸 |	55 𝍸	70 |||
11 𝍸 𝍸 |||	26 𝍸 ||	41 𝍸 |	56 𝍸 𝍸	71 𝍸 ||||
12 ||||	27 𝍸 𝍸 𝍸	42 𝍸 𝍸	57 𝍸 |||	72 𝍸 𝍸 ||
13 𝍸 ||||	28 𝍸 𝍸 |	43 𝍸	58 𝍸 ||||	73 𝍸 |
14 𝍸 𝍸	29 |	44	59 𝍸 𝍸 |	74 𝍸 |
15 𝍸 𝍸 |	30 𝍸	45	60 𝍸 |||	75 𝍸 |
Total 100	100	80	100	100

will put you on a par with the judgment I acquired over decades of experience.

The obvious procedure is to examine the other columns. The second column is more uneven that the first. Three numbers are missing, and 29 appears only once in the 20 cards. Column 3 is not much better. The entries in the last two columns do not depart excessively from the predicted. On the average each number should appear 100/15 = 6 2/3 times. (The numbers must appear a whole number of times, of course.)

Remember that I had started out with the theory that the five numbers found in a column were picked by lot from the 15 numbers available for that column. Examination of these cards makes me doubt this assumed theory. Why? Because I can compute the chance of finding such excessive departures from the prediction. The chance, in this instance, is extremely small. I do take a slight risk in saying that the assignment of the numbers was not done solely by drawing lots. The risk is small. Every scientist when he interprets his data should be able to calculate the risk of drawing a wrong conclusion. We cannot say *how* the numbers were assigned. All we have achieved is our conclusion that they were not assigned solely by lot.

Let us look in particular at the B column on the first of the two cards below. This column has the numbers 11, 14, 6, 15 and 7. Imagine that we now draw, by lot, five numbers from the 15 available for that column. We have already calculated that there are 3003 different sets of five numbers. The column is one of these 3003. The chance that the five numbers just drawn match our column is one in 3003. Sometimes there will be four numbers in common, sometimes three in common, or two, or one or none in common. There is no difficulty in calculating the chance for each of these possible outcomes. We will use the same formula that was used to get 3003 plus just a little thinking.

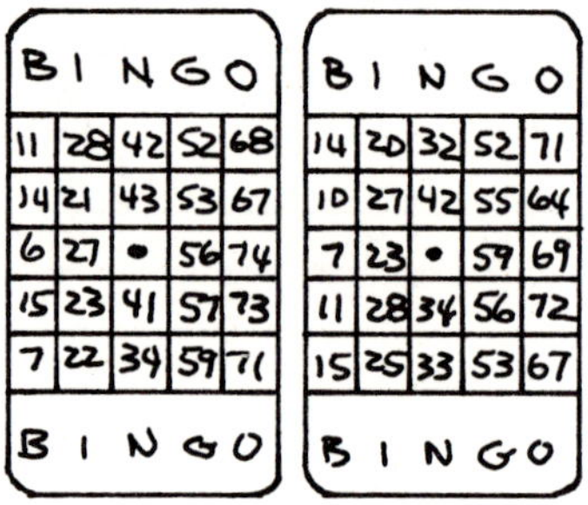

We will do the easy one first and find out how many of the 3003 have no numbers in common with our card column. For this to happen the five drawn by lot must come from the ten numbers ***not*** on our card. Well, in how many ways could you draw five objects from ten? Our formula is:

$$\frac{10 \times 9 \times 8 \times 7 \times 6}{1 \times 2 \times 3 \times 4 \times 5} = 252$$

So there are 252 sets of five numbers that have no number in common with our column.

Now tackle the problem of sets of five that will have one, and only one number in common with the column. One of the five must match some one of the five in our column. The other four must be drawn from the ten ***not*** in our column. These four can be drawn in $(10 \times 9 \times 8 \times 7)/(1 \times 2 \times 3 \times 4) = 210$ ways. The remaining number has to come from the five in the column. One can be picked from five in $5/1 = 5$ ways. And each of these five could be combined with each of the 210 sets of four. The total of such "one in common" possibilities is 5 X 210 = 1050.

For a random set of five numbers to have exactly two numbers in common with our set of five, we must have three numbers that are not in our set. Three numbers can be selected from the ten not in our column, in (10 X 9 X 8)/(1 X 2 X 3) = 120 ways. The two that match our card can be drawn from our five in (5 X 4)/(1 X 2) = 10 different ways. Thus there are 10 X 120 = 1200 ways to have two numbers in common with our five. See if you can figure out the answers for three, four and five in common that follow.

Numbers common to column and sets drawn by lot		Count of possible sets	Percent
None in common		252	8.4
One in common		1050	35.0
Two in common		1200	40.0
Three in common		450	15.0
Four in common		50	1.7
Five in common		1	—
	Total:	3003	100.1

It is most satisfying to see that the sum for all six possibilities is 3003. We have not left out any sets or counted any twice. The agreement of the total with the 3003 obtained by a different calculation verifies our thinking and our arithmetic. There is something inherently elegant in the way the pieces go together. Mathematicians experience this pleasure frequently.

We have made some predictions using a mathematical model. How can we test these predictions using our 20 cards? Here is what I did. I took the B column on the first card and compared it with the B columns on the other 19 cards. For example, looking at the B columns of the cards on p. 26, we see that the first two cards have one number in common (15). The first and third cards also have one number the same (2) as do the first and fourth cards (2). The first and fifth cards have no numbers in common, while the first and sixth have two numbers in common (10 and 15). I made a record of

AGREEMENT BETWEEN COLUMNS

Number Same	B 1-15	I 16-30	N	G 46-60	O 61-75	Total Found	Prediction
0	9	0	-	21	19	49	64
1	49	24	-	61	58	192	266
2	82	81	-	70	72	305	304
3	37	70	-	33	34	174	114
4	13	15	-	5	7	40	13
Total	190	190	-	190	190	760	761

the numbers that were the same for each comparison, recording either 0, 1, 2, 3 or 4. I then compared the B column on the second card with the B columns on the other 18 cards. I had already compared the first and second cards. This process was continued right down the stack of cards. The last comparison was between the 19th and 20th cards. Altogether there were 190 comparisons. The results are shown in the table above. Note particularly that there are 13 cases showing agreement for four of the numbers. This amount is considerably more than the 3.2 mathematics predicts (1.7 percent of 190).

The same procedure was followed for all columns except the N column which has only four numbers. Each column shows an excess of "four in common." The total for the four columns is 40, and this is far too large a departure from the predicted 13. There is also a consistent deficiency in the "one in common" counts. Beyond all reasonable doubt the cards are not what one would expect if chance alone operated in making them.

I can venture a guess about these Bingo cards. There is a little time between the draws of the number—enough time for players to observe the markers on their neighbors' cards. When a player sees a column with four markers on both his and his neighbor's card, it will add considerably to the suspense and excitement. So my guess (and it is only a guess) is that the designer of the cards deliberately made such "four in common" pairs more often. Evidently, he did the same for the "three in common" category. This seems to have been accomplished by sacrificing a lot of the "one in common" pairs. The total lack of cards on which all five numbers in a column agree might be due to chance since it would occur only once in 3003 pairs. It might also have been deliberate on the part of the designer who might have wanted to avoid disappointing the players. This example has shown you, I hope, that an understanding of these matters enables you to "see" things that might not be at all apparent to the uninformed.

Naturally, I was curious to know if all Bingo sets showed the same departure from what we would predict if the numbers in the columns were random selections. I bought two more sets in Washington, D.C. The first, containing 33 cards, appeared to verify the features I found in the Norway set. After a while I discovered that there was a good reason for this similarity. Twenty of the 33 cards were identical with the 20 cards in the Norway set. But the second American set had 20 cards and duplicated none of the cards in the other two sets. What really surprised me was that this set of cards appeared quite free of tampering. I even put them through another kind of test.

The five numbers in a column may form one of seven patterns. All five could be in sequence, as 4, 5, 6, 7, 8. Or there might be a run of three and a run of two, such as 2, 3, 4 and 8, 9. There might be no runs at all; for example, numbers 2, 5, 9, 11 and 14. Now I divided our 3003 possible columns into seven groups. I listed the prediction and the results obtained with the second American set:

Group pattern	Prediction		Found
Run of five	11	.29	1
Run of four	110	2.9	3
Run of three plus run of two	110	2.0	3
Run of three	495	13.2	14
Two runs of two	495	13.2	13
One run of two	1320	35.2	32
No runs at all	462	12.3	14
Total:	3003	80.0	80

These values listed under "***Found***" agree very well with the predicted results, again indicating that this set may have been chosen at random. And, by the way, when you are able to work out the above predicted values you will demonstrate a good mastery of these matters. For example, when drawing three numbers from the set 1, 2, 3, 4, 5, there are ten possible patterns, 3 having a run of three in a row, 6 having two in a row and 1 having none. Do you think you could develop a formula to predict this? Will it work for drawing three numbers from the set 1, 2, 3, 4, 5, 6 where the answers are 4, 12 and 4?

The penny-tossing experiment

4

Some years ago I read a piece in the ***Washington Post*** about a high school student. I quote from the story printed on November 27, 1965. "Edward got 464 more heads than tails and so discovered that the United States Mint produces tail-heavy pennies." You can guess that this item was one that would interest me. The only information in the paper was that 17,950 coin flips had been made, along with the name and address of the young experimenter.

A short bit of arithmetic showed that the excess of heads was far beyond what we predict should reasonably happen with good, unbiased coins. I telephoned Edward and found that he knew something about the mathematics which is the subject of this book. I then asked him how he had done his experiment. For convenience, really to save time and effort, he had tossed five coins at a time. A younger brother had recorded the results as Edward called them out. There are six possible results for a toss: all heads, four heads, three heads, two heads, one head and no heads. Here is the record I obtained from the young experimenter:

Number of heads in toss		Number of times found	Total Heads found
0		100	0
1		524	524
2		1080	2160
3		1126	3378
4		655	2620
5		105	525
	Total:	3590 Tosses	9270 Heads

The total number of tosses is 3590. There were five coins per toss so the total number for coins is 5 X 3590 or 17,950. The mathematical model is based on the presumption that heads will appear as often as tails. The predicted number for heads and for tails is therefore one half of the total number of coins that may have been thrown; that is, the predicted number is half of 17,950 or 8975 heads and 8975 tails.

	Predicted	Experiment
Heads	8975	9207
Tails	8975	8743
Total:	17950	17950

Excess of heads over tails: 9207 – 8743 = 464

One test of the data had not occurred to the experimenter. On the assumption that the coins are unbiased it is easy to predict how often each of the six possible results will appear. Consider the five coins put down in a row with all heads showing. There is only one way to do this. If we want to have four heads and one tail, we must turn over one of the coins. We may turn any one of the five coins; thus there are five ways to get the results four heads and one tail. To leave three heads we must turn over two coins. This can be done in ten different ways. The first coin in the row can be turned over with any of the other four to give four ways; the second coin in the row may be turned over with any of the three to its right to give three more ways; the third coin may be paired with either the fourth or fifth to give two more ways; and finally, the fourth and fifth coins may be paired. Altogether there are 4 + 3 + 2 + 1 = 10 different ways to turn over two coins. Every time two coins are turned over three coins remain. Consequently, there are the same number of ways to select three from five. Continue this approach and you will end up with the prediction that follows at the top of the next page. The 32 possible combinations for five coins are also shown on that page. Thus in one set out of 32 we would expect to find no heads.

Edward made 3590 tosses, so it follows that we would predict 3590/32 = 112.2 tosses with 0 heads. Similar calculations lead to the table at the top of p. 36.

Number of ways to get:	Prediction
0 heads	1
1 head	5
2 heads	10
3 heads	10
4 heads	5
5 heads	1
Total:	32

We can now compare the prediction for each of the six possible outcomes with the results reported by the experimenter. I think the comparison is most interesting. Two areas raise questions. First

POSSIBLE COMBINATIONS FOR FIVE COINS

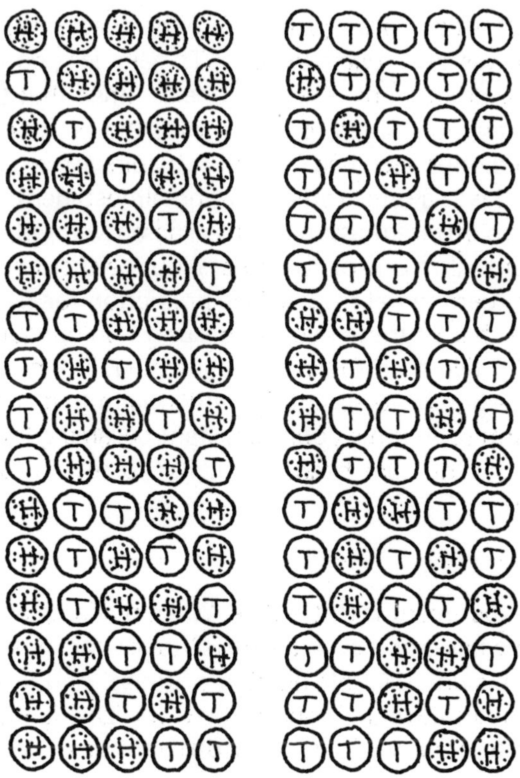

Number of heads	Predicted by theory	Found by experiment
0	112.2	100
1	560.9	524
2	1121.9	1080
3	1121.9	1126
4	560.9	655
5	112.2	105
Total tosses:	3590	3590

look at the entries for four heads. The experiment column shows an excess of 94 tosses over prediction. At four heads per toss this accounts for 4 X 94 or 376 of the surplus 464 heads. It seems most strange that coins with a tendency to fall heads up do not show an excess of tosses with five heads. Instead the number is a few less than the prediction.

The excess of heads does logically raise the question: Are *all* the coins biased or was there ***one*** bad coin? To gain light on this question I made two new predictions. The first one assumed that there was one coin badly enough biased to account for the excess of 464 heads. The second prediction assumed that all five coins were equally slightly biased. I was a bit surprised to find almost identical predictions for the two different assumptions. These two predictions agreed with each other within one or two tosses, but neither lined up at all well with the experimental results. The prediction for five heads was 127 tosses by either of these two assumptions. So our surprise at getting only 105 five-head tosses seems justified. Taking all this into account, I am reluctant to accept Edward's claim that cents are tail-heavy. I surmise that his young assistant became confused and recorded many of the "one head, four tails" tosses as "four heads, one tail" tosses.

The above experiment is instructive. When planning an experiment try to find some basic unit such as the 32 tosses for the coin experiment. Repeat this unit a whole number of times. And introduce, if you can, some logical subdivisions in the work. Divide the work into morning and afternoon sessions, or by different days. The various sections of work can then be examined to see how consistent they are. For example, the coin experiment as performed could not throw any light on the question of whether one coin was faulty or all were biased. The coins could have been chosen with different dates. If years ending in 5, 6, 7, 8 and 9 were selected, it would suffice to record, for each toss, the number on each coin showing a head. There would be much less likelihood for the error of confusing one-four tosses with four-one tosses, and much more information per toss. If the coins really were biased, the separate data for each coin would carry conviction. A bit of extra work might be required, but it would probably be worth the effort.

There are other intriguing possibilities. Suppose we selected the following four coins:

Denver mint	1957, 1968
Philadelphia	1956, 1969

You can tell coins from the Denver mint by a small "D" near the date and San Francisco coins by an "S." Philadelphia coins have no letter. Now flip the four coins and record which ones come down heads. Record your results in a table such as below.

	Number of heads		
	Old	New	Total
Denver Mint	A*	C	A + C
Philadelphia Mint	B	D	B + D
Totals:	A + B	C + D	A + B + C + D

*Replace the letters A, B, C and D by actual counts.

The advantages of this scheme are pretty obvious. For the same amount of work we can compare the mints by making use of the row totals; and we can compare old and new coins using the column totals. If it should turn out that there is a difference between coins from the two mints, we can see whether this difference remained the same over the years. Thus A – B gives the difference between mints in the fifties, and C – D gives the difference between mints in the sixties. If we want to compare these two differences, we can subtract one from the other: (A–B) – (C–D) = (A + D) – (B + C).

So we have not only (A + B) - (C + D) and (A + C) - (B + D), but this third way to examine the data.

Exactly this scheme, and extensions of it, have been used in countless researches. Of course, not everyone is convinced. Some are frightened by mathematics and others have never perceived the beauty and efficiency of performing experiments in this way. The really important lesson here is: ***Do not rush into your experiment***. First spend a little time seeing how you can increase the amount you can learn from the same amount of experimental work.

Experiments with cards

5

Working with cards will introduce an idea not mentioned before. First, let's return to Chapter 2 and the marble experiment. By direct enumeration we were able to predict that all four first places would be occupied by red marbles in one of the 70 orders the marbles might give. We might, instead of using the trough to establish an order, simply draw marbles, one at a time, from a box and record the color.

We may easily calculate our chances of getting four red marbles from a box containing four red and four yellow marbles. For the

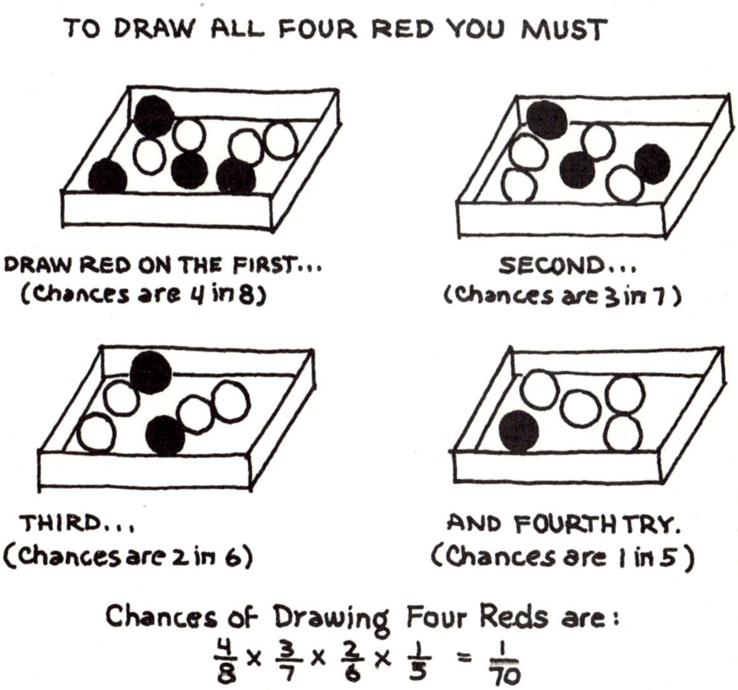

first draw from the box our chance of getting a red is one half because there are equal numbers of both colors in the box. Put the marble aside and draw a second marble. If the first was red, there are now three red left in the box and four yellow for a total of seven marbles. The chance of now getting a red one is three out of seven or 3/7. If it is red, then on the next draw the chance is two out of six or 1/3. Finally, when three red have been drawn, the chance of getting the last red marble is one out of five or 1/5. When we combine these four draws we have $1/2 \times 3/7 \times 1/3 \times 1/5 = 1/70$. By a different approach we have now verified the one in seventy chance that we obtained in Chapter 2.

In Chapter 2 we also counted the number of red marbles in the first four places after each trial with the trough. We can go back to the same data record on p. 9 and examine the first two places in the trough. I have counted these and found the following results for the 70 trials.

	Found	Prediction
Red marble in neither position 1 nor position 2	16 times	15
Red marble in position 1, not in position 2	22 times	20
Red marble in position 2, not in position 1	20 times	20
Red marbles in both positions	12 times	15
Total:	70	70

What would we predict? For the first line $1/2 \times 3/7 = 3/14$ is the chance. Why? When 70 trials are made we would predict $3/14 \times 70 = 15$ times. The 16 found is very close. For the second line we would predict $1/2 \times 4/7 = 4/14$. For 70 trials this leads to the prediction $4/14 \times 70 = 20$. Symmetry leads to the predictions for the last two lines. In this case the found values were acceptably close to the predicted.

But suppose there are only two red and two yellow marbles in the box. What would we then predict for the first two positions? To get two red we follow the above reasoning and multiply $1/2 \times 1/3$ to get 1/6. To get a red and then a yellow we multiply $1/2 \times 2/3$ and get 1/3. Similarly, the chance for a yellow followed by a red is also 1/3. And to get two yellows it is the same as for two reds. These four fractions obtained with four in the box can be compared with the four fractions when we start with eight in the box.

	Red Red	Red Yellow	Yellow Red	Yellow Yellow
Four in the box	1/6	1/3	1/3	1/6
Eight in the box	3/14	4/14	4/14	3/14

There is quite a difference between 1/6 and 3/14. In decimals it is 0.17 and 0.21. The important point here is that a marble, once drawn, is not put back in the box. But if each marble is replaced (after noting its color) the prediction will be the same for four, or eight, or eight-hundred marbles.

When a hand of cards is dealt from a pack of cards we are involved in a situation of ***nonreplacement***. Consequently, our predictions for card hands must take account of this fact.

Suppose we put four red and four black cards together and shuffle them well. If four cards are drawn, the chance of getting four red cards is, as we have seen above, $4/8 \times 3/7 \times 2/6 \times 1/5 = 1/70$. The usual deck contains 52 cards, half red and half black. If four cards are drawn (without replacement) from a full pack, the chance that they will all be red is $26/52 \times 25/51 \times 24/50 \times 23/49 = 46/833 = 0.055222$. This is nearly four times the $1/70 = 0.014286$ that applies to a pack of eight cards. Should four ordinary decks be combined to give a pack of 208 cards, the chance becomes $104/208 \times 103/207 \times 102/206 \times 101/205 = 1717/28290 = 0.060693$, and we are approaching a limit which would be given by packs of infinite size. For such a pack the withdrawal of a finite number of cards would not alter the ratio of red cards to all the cards in the pack. The ratio would remain 1/2. It follows, therefore, that the chance of drawing four cards all red is $1/2 \times 1/2 \times 1/2 \times 1/2 = 1/16 = 0.062500$.

As a matter of fact we frequently realize this situation. If a fair coin is tossed once, the chance of getting a head is 1/2. When the same coin is tossed repeatedly, the chance of getting a head does not change—the coin has no memory. Consequently, the chance of four tosses giving four heads is 1/2 X 1/2 X 1/2 X 1/2 = 1/16. This arithmetic can easily be verified by enumerating all 16 ways in which four tosses of a coin may appear. We did that in the preceding chapter for five pennies.

Now let's do an experiment. The first requirement is to make sure that the cards are properly shuffled. Many card games introduce a grouping by card suits or sequences in denomination. Casual shuffling will not put the cards into a random order. My practice in shuffling cards is to deal out the cards into six or more heaps. I skip heaps and even vary the order in which the heaps get each additional card. I reassemble the pack and repeat this operation twice more. In our work we will need to do this only at the start. We will follow a routine that will not introduce any pattern in the sequence of cards in the pack.

In this experiment I used the following method for selecting eight cards. The properly shuffled pack was held with card backs up. The top cards were slightly spread, and at an arbitrary point the top few were removed and then placed in a heap face up. Only the top card showed. The number of cards so removed varied from two to ten, on the average five or six. This was repeated seven times more, all piles being placed face up. The hand of eight was thus revealed and a count made of all red cards, R, and all face cards, F.C.

Two experiments were performed, each of 100 hands. The complete data are shown to the right. The data for each experiment consist of five columns; each column consists of 20 pairs of integers. The first digit in each pair refers to the number of red cards found in the hand of eight cards. The second digit refers to the number of face cards in the *same* hand. The number of red cards can vary from zero to eight. The number of face cards can also vary from zero to eight. Only once were as many as six face cards found in a hand.

A count was made of how often each possible outcome occurred for red cards and for face cards. These counts are shown on p. 44. Following our usual approach I made predictions for each case, and these are shown in the two columns to the right. The data for red cards exhibit the same symmetry as the predicted values. This symmetry is to be expected, because there are equal numbers of red and black cards in the pack. The number of black cards in each

DATA FOR TWO EXPERIMENTS: EACH OF 100 EIGHT-CARD HANDS

Experiment 1

40	52	30	53	42
51	40	71	33	33
43	33	70	42	43
51	73	52	60	51
40	41	52	61	32
35	30	41	41	34
41	41	30	52	40
64	21	32	42	52
62	31	50	33	61
64	52	62	63	24
51	42	32	62	41
42	50	33	23	52
43	51	44	42	62
12	51	52	54	32
41	33	22	63	63
42	30	52	52	51
41	61	61	50	63
53	43	22	43	52
34	31	42	33	21
32	31	42	41	42

Experiment 2

23	32	60	42	41
41	51	61	42	31
42	32	44	11	44
42	33	53	73	41
33	22	56	41	50
32	32	34	52	43
21	72	21	42	62
41	61	63	13	42
53	61	40	32	63
43	02	33	41	41
41	61	22	53	33
52	82	63	61	31
41	32	22	42	41
54	52	40	51	40
42	41	01	43	30
42	52	33	51	32
31	41	41	30	52
31	62	53	55	42
41	73	41	40	63
51	51	53	51	41

hand can be found by subtracting from eight the number of red cards recorded for that hand in the data table.

The red card predictions were obtained by following the same reasoning used in Chapters 2 and 3. To get eight red cards we must calculate in how many ways eight red cards can be selected from the available 26 red cards in the pack. To get seven red and one black we calculate all the ways to pick seven red from 26 red and

TABULATION OF HANDS CONTAINING DIFFERENT NUMBERS OF RED CARDS AND FACE CARDS

Number in Hand	Experiment 1		Experiment 2		Predicted	
	RED	FC	RED	FC	RED	FC
0	0	13	2	8	.2	10.2
1	1	26	2	35	2.3	29.8
2	6	33	6	30	9.9	33.7
3	22	20	19	21	22.7	19.2
4	29	7	35	4	29.7	6.0
5	24	1	20	1	22.7	1.0
6	15	0	12	1	9.9	.1
7	3	0	3	0	2.3	.0
8	0	0	1	0	.2	.0
Total	100	100	100	100	99.9	100.0

one black from 26 black and multiply these together. The results of this arithmetic are listed under "Predicted—Red." Can you check these? The same procedure gives the prediction for the face cards. The face cards number only 12 out of the 52 in the pack, so hands of eight would be expected to average around two face cards. The predicted values are shown in the last columns and agree satisfactorily with the experimental values. In this experiment results agree well with predictions, so we will move on with more confidence in our ability to predict and to do card experiments.

Let's stop for a minute, however, and mention disagreement between results and predictions. Using a mathematical model that does not fit the problem might be cause for disagreement. In the case of playing cards, the model we used assumes proper shuffling. Inadequate shuffling could be an explanation of disagreement with prediction. Unsatisfactory clay marbles were responsible for disagreement in Chapter 2. In other cases, disagreement is caused by carelessness in recording the data or making summaries of the data. Probably poor recording was the difficulty in the coin experiment in Chapter 4. In real research the experimenter must always guard against being misled by insufficient randomization, defective apparatus, observer bias or even simple carelessness on the part of assistants.

One interesting example happened to me early in my career. Many studies had been made of how to store seeds so as to keep them for a long time and have them retain their ability to germinate. After the seeds had been stored at various temperatures or with air replaced with other gases, samples would be taken to ascertain the percent of seeds that would still grow. Usually two or more samples of 100 seeds would be placed on filter paper, wet with distilled water, and the two or more dishes placed in a warm chamber to induce germination. When all had germinated that were going to, a count would be made for each dish.

When I studied the records of published research, I found that identical samples generally disagreed more than mathematics predicted. Such extra disagreement between samples of seeds receiving the ***same*** treatment would tend to obscure possible differences between seeds stored under really ***different*** conditions. To check my mathematics I did a simple experiment. I took a supply of seeds and colored a portion of them with a dye. The colored ones would play the role of non-germinating seeds. I mixed colored and uncolored seeds carefully and withdrew samples of 100 without allowing color to influence me. I used a small cup that held a few less than 100. Any additional seeds were withdrawn one by one without looking in the jar. Then I counted the colored seeds in these samples. These counts did check prediction, giving me added confidence in my mathematics. I was able to conclude that somewhere in the very simple routine followed to get seeds to germinate there must be a lack of control. That is, some source of variation was present that caused the extra variation found in seed studies. After some investigation, I discovered that the incubating chamber, which was large enough to accommodate many samples, was not uniformly warm all over, a likely source of the trouble.

The real lesson from this aside is that one should never fail to check carefully if the mathematical model really is suited to the phenomena under study. For this reason I formed the custom in my researches of including extra items that were identical to some of those under study. Often it was easy to conceal from assistants that two items which had different labels were actually identical. There is a universal tendency to make the measurements agree, if it is known that the items under measurement are the same. Certainly it would be unwise to be impressed by an observed difference between two drugs, if it was no greater than that reported between two samples of the ***same*** drug. Such concealed duplication does make it possible to test the adequacy of the assumed mathematical model. And, best of all, it protects the researcher from making claims that he may later have to withdraw.

Beginning more than 40 years ago, some very serious studies were undertaken to determine if information could be gained or transmitted by means other than the five senses that we all use to obtain an awareness of the world around us. The research was an attempt to find out whether some people could acquire information, when all known means of getting that information were carefully eliminated. Since none of the usual senses are involved, these studies investigate "extrasensory perception," which is usually shortened to "ESP." ESP research was supported by grants of money, a department was formed at a large university, and a journal was established to publish the results of research performed in this field.

I discuss this subject here because one of the major tools used in these researches was a specially designed pack of 25 cards. The cards were about the size of ordinary playing cards, printed in black. There were five different symbols: a circle, a square, a star, three wavy lines and a cross. There were five cards having each symbol—only one symbol on a card. The cards were used in many ways. For instance, a well-shuffled pack would be taken by the experimenter, who, concealed by a screen, would go through the pack, look at each card, and wait for the subject to make his guess. A record was made

THE FIVE SYMBOLS USED IN ESP CARD PACKS

of the symbols on the cards and the corresponding guesses. A count was then made of the number of correct guesses, or hits. Prediction for hits, if guesses are made at random, is five hits in a pack of 25. Exactly five hits will be obtained if the subject calls the same symbol every time. One unforeseen complication was that subjects rarely divided their guesses equally among the symbols.

There were frequent reports of unusually long runs of hits—runs that were beyond a prediction based on chance. Many variations of this card game were tried. Sometimes experimenter and subject were in different buildings. Sometimes the experimenter did not look at the cards until after the subject had made all 25 of his guesses. Even more extraordinary, in some trials the subject made his 25 guesses, the cards were ***then*** shuffled, and the newly shuffled pack compared with the guesses. Getting more than the predicted five hits was considered ***precognition***, that is being able to foretell the order that the cards would have ***after*** being shuffled. Sometimes the cards were not removed from the pack, but the guesses would be made without disturbing the pack until after the guesses were recorded.

It is little wonder that claims made for hits substantially in excess of prediction were generally greeted with skepticism. Critics felt that there must be some explanation other than this unknown power of extrasensory perception. As accusations of cheating were made, more and more precautions were taken to avoid giving the subject any clues. At the same time reports of success were made by workers in other laboratories. There were also reports of failure. The claims made by believers were after all fairly modest. Only a small fraction of the population had the power, and then usually only to a minor degree. The power varied from time to time and often eventually was lost completely.

The disbelievers were not impressed by averages of many trials; quite often these averages were only around six or seven instead of the predicted five. The very large number of trials did, according to the believers, reduce the risk of purely chance results to one in billions. This field of investigation posed worrisome problems to the physical scientist. His goal is to remove the observer from the phenomena under study. Should a chemist discover how to make a new compound, he must then describe his method so carefully that ***any chemist, anywhere, at any time*** can make this same product. Men whose lives have been guided by such requirements are apt to be impatient with so highly personal a matter as ESP. This reaction is especially evident when the proponents claim that a skeptical audience may put the subject off his form. Scientists who would not dream of putting a pad down before a poet and demanding a poem on the spot, or a composition from a musician, demanded that the subject demonstrate his power when challenged. The scientist is used to meeting such demands. The advocates for the existence of ESP were so anxious to furnish "scientific proof" that they forgot that the rules for scientific evidence are very demanding.

When I seek an explanation for results beyond those predicted, I am bound to consider various ways in which this could happen as a result of faulty experimental technique. One common source of error is to ***select*** a portion of the subject's record ***after*** going over the data. Another error is to terminate the experiment when the experimenter feels that the subject is beginning to lose his power. Still a third trap is to keep in the record a very promising series of results obtained by a subject during a preliminary investigation to select the best performer from a large group. All three of these practices destroy the validity of the comparison with prediction. The mathematics of prediction permits no selection based on a study of the data after it is obtained. I suspect that much ESP work has errors of this nature. Scientists want more information on questions such as What conditions are favorable for ESP power? and What guides are used to determine the selection of good subjects?

The whole subject of ESP runs contrary to long-established attitudes. For example, suppose a man and his brother leave a house together. The man returns to get a forgotten item. As he comes out he sees his brother attacked by a red-haired man who limps and makes his getaway in a black car. He immediately calls the police, who catch a suspect meeting this description. The man identifies the suspect, and his *visual* evidence gets a conviction in court.

Suppose the man's brother were in another city when he was attacked. The man has a dream exactly like the scene described above. It is highly improbable that if he called the police in the distant city, the police would use the dream as a guide in the search for the attacker. Suppose the police learn that the brother was seen having an argument with a man which they subsequently locate, and he has red hair, a limp and a black car, just as in the dream. There is no other evidence aside from this dream, and the brother never saw his attacker. No jury would hand down a conviction on the basis of this dream. Probably, the jury would not even be allowed to hear about the dream. So we come hard up against a long-established standard of what constitutes evidence.

The subject of extrasensory perception will continue to attract attention, and will certainly remain controversial for a long time. My own point of view is that much of the experimentation has been poorly planned. The ESP cards, designed 40 years ago, should long ago have been replaced. The usual playing cards seem to me to be more useful, for they make it possible to see if color is a factor, and certainly the face cards are a sharp contrast to the cards with spots.

Once during a convalescence, I collected a lot of data using playing cards. I removed the four 10-spot cards from a deck to reduce the pack of cards to 48. As there are 12 picture cards, I had now made the proportion one-fourth, thereby making the arithmetic a lot easier. The task I set myself was to select a hand of four cards from the deck with the avowed intention of doing better than one face card per hand. In some ways this is simpler than the ESP cards with their five not too dissimilar symbols. Surely, I thought, this power of ESP should find it much easier to tell face cards from cards with spots than to identify the ESP symbols. I went through the pack, card by card, setting aside any card that I "felt" was a face card. I stopped as soon as I had removed four cards. I then made a record of these cards. It is easy to calculate predictions of zero, one, two, three or four face cards on the assumption of no ability. My performance was right in line with prediction, so there was no evidence of any ESP talent on my part.

I continued my experiment. I changed my method of selecting four cards. I shuffled carefully, cut the pack and noted the number of spots. All face cards received the value of ten. From this point on I had no further control of the selection. I used this number to count down in the pack to determine a stopping point. This portion was removed and set aside face up. The number of this card was used to count down and remove a second portion also placed face up. The process was continued until four portions were removed and I could make a record of the four cards showing. This constituted a trial. The number on the last card in a trial was used to start the next selection of four cards. The sections were reassembled in a different order and the pack given a slight shuffle before starting with the number shown by the last card in the preceding trial. Thus I had an automatic selection process determined by the cards themselves.

I performed the following experiment. The schedule that follows shows that I (as Mr. Y) made ten trials in the morning. I then changed myself to Mr. X and made another ten trials. The program of 20 trials was repeated in the afternoon. This whole schedule was followed for four days making a total of 160 trials with a predicted total for face cards also of 160.

	Morning		Afternoon	
Monday	Mr. Y	Mr. X	Mr. Y	Mr. X
First set of 5 trials	—	—	—	—
	—	—	—	—
	—	—	—	—
	—	—	—	—
	—	—	—	—
Second set of 5 trials	—	—	—	—
	—	—	—	—
	—	—	—	—
	—	—	—	—
	—	—	—	—

Repeat above schedule on Tuesday, Wednesday and Thursday.

What could I do with all these results? Really, quite a lot. Mr. Y and the mythical Mr. X had each made half of the trials, so the face cards should be equally divided. Half the trials were first sets and half second sets, so again the face cards should be equally divided. In an actual experiment we would be looking for a fatigue effect. Half of the work was done on Monday and Tuesday and half on Wednesday and Thursday. Again, I looked to see how well the two halves agreed in the face cards. Also, I counted up all red face cards and all black face cards. If one color was more easily detected we would look for an effect here. It was easy to compare the performance of Mr. Y and Mr. X for kings, queens and jacks.

	Number found		
	Kings	Queens	Jacks
Mr. Y	—	—	—
Mr. X	—	—	—
Totals:	—	—	—

The predicted number in each cell would be 160/6 or 26 2/3. We also know how to test for departures from prediction. Notice how many different questions we are able to ask our data. Of course this requires some planning, adherence to the plan, complete avoidance of selection, and avoidance of the temptation to introduce some spontaneous variation in technique. Such planned experiments are a great improvement and permit the investigation of several questions simultaneously. Great efficiency is needed to establish small differences and such planning has not been evident in ESP work.

Almost all experimental work with plants and animals makes use of "controls." A new drug is pitted against the old drug which serves as a control. In ESP work the subject is pitted against mathematics. But we have seen earlier in this chapter that the extremely simple experiments on seed germination did not conform to the mathematical model. It would be interesting to have a subject who was picked as possibly having ESP power paired with a "control" who has never exhibited such ability. Both persons should be tested simultaneously with the same pack. A good idea would be for the experimenter (or recorder) not to know who was the control. Nor should the one examining the data! In such a manner we would satisfy ourselves that the mathematical model did apply to the technique, cards, and circumstances that existed during the test.

Experiments with dice

6

In another set of experiments I used five different dice, which could easily be identified by color or size. One die was red (R), one green (G), and one orange (O), all bought in Washington, D.C. In Norway I bought a little (L) white die and a bigger (B) white die. In these experiments it will be most important to be able to follow the performance of each die.

If you have some dice, I suggest that you examine them carefully. You may have noticed that the sum of the dots on opposite faces of a die is always seven. Try placing a die with the 3 uppermost. The die then rests on the side with four dots. Now turn the die, keeping the 3 face on top, until you see the sides with one and two dots. A glance at the sketch on p. 54 will show you that all dice are not alike. The large white die has the 2- and the 1-dot faces in reverse order from the other dice. I was quite surprised because I had never previously observed any such variation among dice. Nor, apparently, has anyone else that I have talked to. The authority on these matters is ***Scarne on Dice***.* In his book you will find a virtual treatise on dice, including many different games, the correct odds, and an entertaining account of the endless varieties of trickery that have been devised.

I named the bigger white die a "left-handed" die. The others are right-handed dice. So far I have turned up only three left-handed dice. Scarne does not mention this possibility. He does, however, reproduce a picture, which also appears in the Encyclopedia Brittanica, of an Egyptian and two Roman dice. The Egyptian die is a left-handed one. In Scarne's book there are many pictures of dice

****Scarne on Dice***, John Scarne, Stackpole, Harrisburg, Pa., (1962).

RIGHT- AND LEFT-HANDED DICE

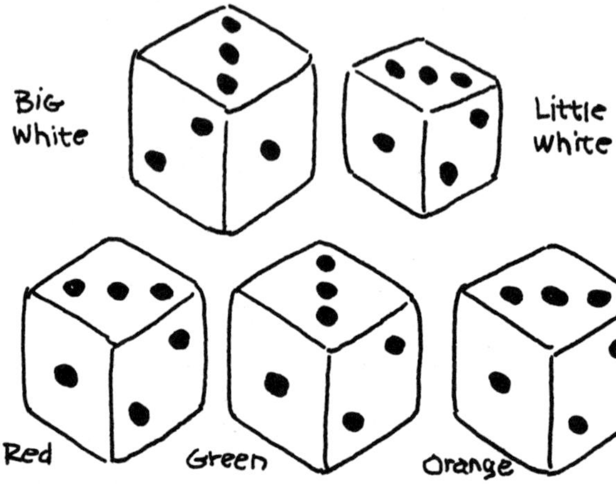

which show three faces, and consequently, you can see whether or not they are right- or left-handed. On page 297 Scarne is showing how to substitute crooked dice for good dice. One of the good dice appears to be left-handed. The next page shows how the crooked dice fall and both are left-handed.

Scarne does mention another deviation that I also found exhibited by my dice. You may have noticed in my picture that the green die has the three dots placed differently from the other right-handed dice. Scarne says this also is done with the 2-spot face to enable cheats to identify quickly their own crooked dice. But for my dice I think this variation was accidental. I had a pair of the green dice, and the other one has the 3 in the regular position. You can't see it, but the green die has the six dots in a horizontal position. The other four dice have the six dots arranged vertically. Altogether then, there are 2 X 2 X 2 X 2 = 16 different recognizable types of dice. Both the left- and right-handed dice may have the two spots, the three spots, or the six spots placed in either of two positions. You might try writing out the possibilities.

If you consider the central point of a perfect cube, you will see that all six faces are equally placed in respect to this center point. Intuitively, one feels that the symmetry of a die gives no advantage to any one face when a die is thrown. This being so, why should we spend time doing experiments with dice? Mainly to gain some acquaintance with the laws that govern variation. The lessons learned from these experiments are a great help in the interpretation of important data.

Incidentally, the cube is one of the five completely regular solids (the Platonic solids). The largest, the isocahedron, has 20 faces, all identical, equilateral triangles. In Japan you can buy sets of three of different colors with the ten digits occurring twice on the 20 faces. They are useful in obtaining random numbers. Another useful device is a ten-sided cylinder, three centimeters long and about the same width. On the ten sides, near one end, the ten digits are stamped in a random order. The object and its use are the work of Dr. H.C. Hamaker, an internationally known statistician from the Netherlands. The cylinder is not rolled on a flat surface as you might guess. Rather it is tossed into the air, imparting a spin around the cylinder axis. As it descends it is caught, and the thumb is pressed against one of the sides ***before*** looking at the cylinder. If it were rolled we should need a more perfect cylinder to avoid bias.

In our experiment we will roll a die six times and keep a careful record of the face that shows on each throw. The results of 12 trials, six with a red die and six with a green die, are shown on p. 56. The first trial gave the face 2 twice, the face 3 twice, and faces 5 and 6, once. Faces 1 and 4 did not appear. This representation is more vivid than a simple written record. The six trials with the red die were as follows: 223563; 624236; 235522; 231653; 651352; 531245, in the order in which the faces appeared. Sometimes the order is important.

After a look at the 12 trials you may conclude that it is impossible to make any prediction about the next trial. I agree that any arrangement of six squares that you might make in advance has a very poor chance of turning up. But all is not lost. I can make at least one prediction with a very good chance of being correct. I notice that in all 12 trials at least one of the numbers is missing, so I venture that in the next trial at least one face will be missing. I did 100 such trials, and every set had at least one face missing. I can add to my prediction the further statement that I would not expect the next trial to show ***more*** than three missing. In other words, I venture to say that in another trial you will get at least three, but not more than five different faces. This prediction turned out to be true for all of my 100 trials. But, as you will see later, I should have expected about four exceptions in 100 trials.

TWELVE EXAMPLES OF SIX THROWS OF A DIE

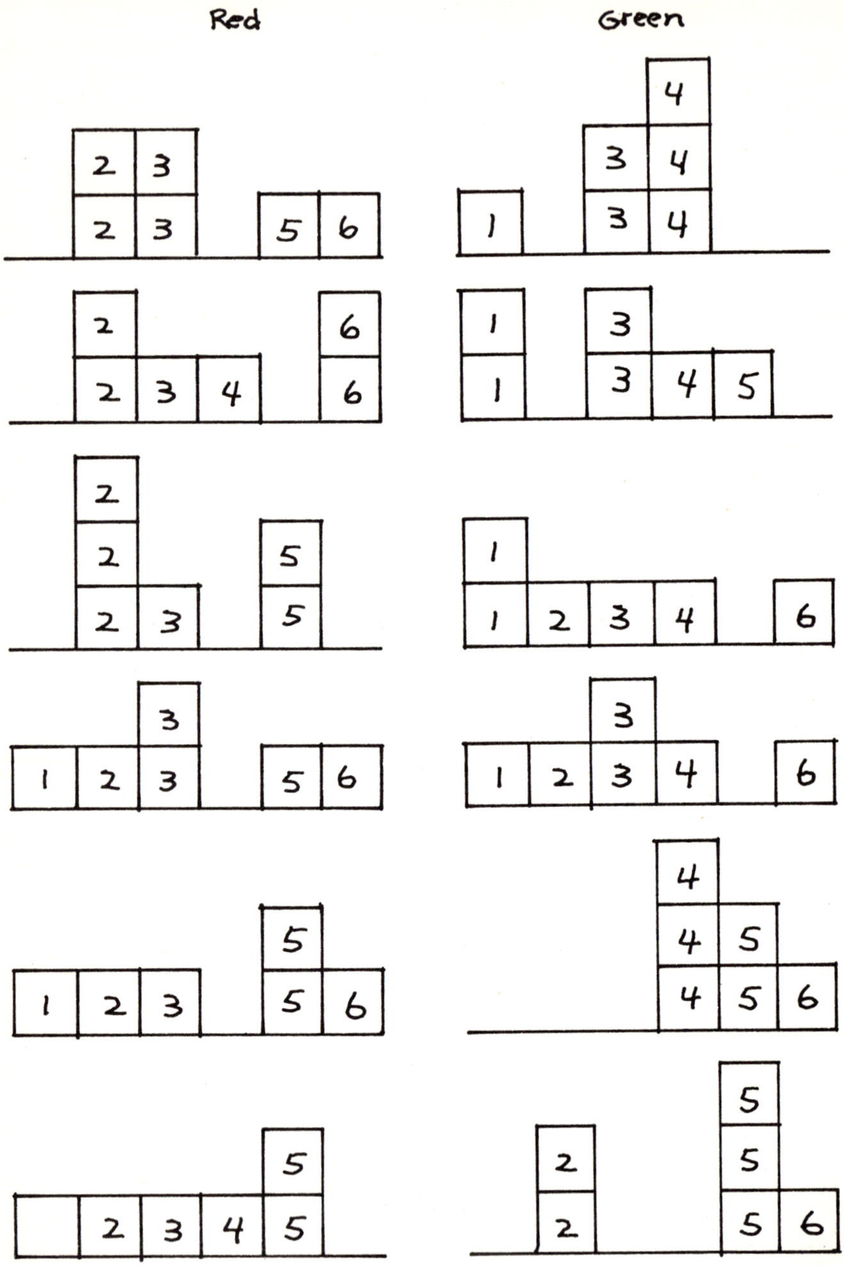

Can we, on the basis of a dozen trials, say anything about the relative frequency of sets showing three, four or five faces? Maybe a little, but there are not enough data to give a reliable answer. You will observe that among the dozen sets, no two are exactly alike. And no wonder, because there are 462 patterns in which a set of six dice can fall.

Where did the number 462 come from? I certainly did not write down every one of the 462 patterns. Not only would this be quite a task, but unless some very good system of enumeration was devised, you would soon be lost in confusion. Resort was made to rather simple mathematics plus a little helpful thinking.

Take a look at the fourth, fifth and sixth red die patterns. They all agree in a particular respect. Each of these three patterns has two of one number and single appearances of four other numbers. So if we ignore the actual numbers, we can group these three patterns together. This type of pattern has one pair and four singles. How many possible patterns are there in this family? Again, we proceed logically. There will be a group where the same face is always missing. For example, the red fourth and fifth sets are both missing face 4. However, the two sets do differ. In the fifth set it is the 5 that has the pair, and in the fourth set it is the 3 that gives the pair. Clearly there must be three other similar sets where 1, 2 and 6 are the pairs. So that makes five sets altogether where 4 is the missing face in every case.

The same argument can be made if some number other than 4 is missing. We stick to the one pair, four singles situation. We can make five different patterns for each one of the six numbers that may be missing. For example, set six of the red die has the 6 missing. This set is one of the five where, in turn, either the 1, or 2, or 3, or 4 or 5 supplies the pair. We may quickly say that there are five times six or 30 possible sets in this family—the family of all patterns that have one pair and four singles as a result of six throws of a die. You will find these results entered in the second line of the table on top of p. 58.

Hold on a minute. We have been talking about rolling one red die six times. Is that absolutely the same as rolling six dice all at once? Don't take my word for it. Stop and think it out. Of course, we do assume the dice are all the same. We start out with the assumption that our red die is perfect. A badly loaded die might always show a 6. In that event the whole discussion is futile. We will, a bit later,

WHAT MAY HAPPEN WHEN YOU ROLL SIX DICE

Number of Faces (Family)	A Typical Pattern	Number of Patterns	Total Number of Ordered Ways
6	316425	1	720
5	554361	30	10,800
4	442251	90	16,200
4	333425	60	7,200
3	555221	120	7,200
3	444426	60	1,800
3	661144	20	1,800
2	666644	30	450
2	666333	15	300
2	555551	30	180
1	222222	6	6
		462	46,656

examine 108 throws for each of these five dice to see if there is evidence of any loading.

Let us now return to our table of patterns. Obviously, there can be only one where all faces are present and only six possibilities where only one face shows. The last entry in the table, six 2's, is an example of this pattern: six of a kind. The table lists 11 possible families. I safely defy you to find any more. However, these 11 families differ considerably in their frequency of appearance.

The frequency of appearance of the different patterns is related to the number of different *orders* in which the faces may appear when a die is thrown six times. Just on intuitive grounds we can see why this should be so. To get six of a kind, it is fairly obvious that the result of the first throw absolutely determines what must happen in the next five throws. There's only one order to the numbers 222222. In contrast, the pattern 316425 is far more elastic. In fact we don't care which face shows first (that gives us six possibilities); then any one of the five other faces will keep us in the game. Continue this argument and see that the pattern "all six faces different" may occur in 6 X 5 X 4 X 3 X 2 X 1 = 720 different orders when a die is rolled six times.

Let's skip to the next-to-last family, shown as 555551 in the table. How do we know there are 30 patterns? You can pick the two different numbers in 15 different ways. The 15 pairs are 1, 2; 1, 3; 1, 4; 1, 5; 1, 6; 2, 3; 2, 4; 2, 5; 2, 6; 3, 4; 3, 5; 3, 6; 4, 5; 4, 6; and

5, 6. Once you have a pair selected, then either one of the pair may be the one to be repeated five times to make the string of six numbers. For example, 111112 and 122222 can be made from the pair 1, 2. That doubles the 15, making 30 patterns in all. In this case we have enumerated the 15 different ways. A formula that will tell us the number of ways to pick two different numbers from six is (6 X 5)/(1 X 2) = 15. To pick three from six, the formula is (6 X 5 X 4)/(1 X 2 X 3) = 20. You can see by trial that this formula works. At some future time you will probably derive it mathematically. You might be able to use it now to find where the 20, 60, and 120 came from for the families having three faces.

In how many different ***orders*** can such a combination as 555551 appear? By trial you can see that the 1 may appear first or at any place up to last. That is, the 1 may appear in any of six positions. Consequently, we multiply the 30 patterns by six and get the 180 entered in the last column of the table. In a similar way I obtained all the entries in the last column; that is, I multiplied the number of patterns in each family by the number of different orders possible for each pattern. I have a very easy check on my arithmetic. All 11 entries add to 46,656. I know this sum is correct, because each separate throw of a die can give any one of the six faces. Six throws make possible 6 X 6 X 6 X 6 X 6 X 6 = 46,656 for the total number of different ways in which the dice can appear when we keep track of order.

There is something else we can do. Observe that among these 11 families only one family has all faces present. There are two families with four faces present, and so on. This observation invites us to group these 11 families on the basis of the number of different faces present. This I have done in the table on p. 60. For example, we predict that in 100 trials we will get about 100 X 720/46656 = 1.6 results showing all six faces. Comparing the observed values with prediction shows that while my dice rolls didn't come out exactly as predicted, they were close.

This grouping technique is a very useful tool in many fields of endeavor. With a small number of groups, we have some hope of seeing how well our calculated predictions agree with what happened experimentally, whereas a much larger program would be required to check on 462 different patterns that differ from each other as wildly as those shown in the 12 original block diagrams on p. 56.

Number of faces	Ordered ways to occur	Predicted in 100 trials	Found in 100 trials
6	720	1.6	0
5	10800	23.1	28
4	23400	50.2	45
3	10800	23.1	27
2	930	2.0	0
1	6	0.0	0
Totals:	46656	100.0	100

These results are of great importance. Pure chance alone operates in the individual items, yet we are able to make quite satisfactory predictions about a collection of sets—even as small a collection as 100. Here we are at the very heart of the scientific method. We need imagination to formulate a model that we feel approximates the actual situation. Using this model we make some predictions. Then some experiments are performed. If the experiments agree pretty well with the predictions, we consider our model a useful tool to guide more research. Sometimes the model is solely a result of imagination, or sometimes the model is suggested by a careful study of preliminary data.

If the model is a guide to further discoveries, it serves the experimenter well. One of the most famous models was made by the chemist Mendeleff. It appeared to him that some elements were missing from his groupings of the chemical elements. From the chemical properties of the known members of the groups, he was led to predict the approximate properties of the missing individuals. With this information as a guide, a search was made for the previously overlooked elements. Three were found during his lifetime.

Before we part company with our dice, we will take a look at the results obtained with just two dice. In this instance, it is easy to make a complete enumeration of all possible outcomes. For our purpose it helps to be able to tell the dice apart. We will use the red die and green die. Our model assumes all faces have the same chance and also that no matter what face appears on the green die there is no effect whatever on the action of the red die. First, let's take a look at the checkerboard chart on p. 61.The six rows correspond to

SUMS OF TWO DICE

Face on Red Die

Face on Green Die	1	2	3	4	5	6
1	2	3	4	5	6	7
2	3	4	5	6	7	8
3	4	5	6	7	8	9
4	5	6	7	8	9	10
5	6	7	8	9	10	11
6	7	8	9	10	11	12

Sums

the six faces on the green die and the six columns to the six faces on the red die. The intersection of rows and columns gives 36 small cells which account for all possible combinations of the two dice. There is no reason to expect any one of the 36 cells to be favored over the others. We would predict that the results of many throws would be distributed uniformly over the whole checkerboard of possible combinations. But, again, 36 items is a rather large collection, and grouping may be necessary.

We have entered in each cell the sum of the faces shown by the row and the column that intersect to locate the cell. These sums range from two to 12, and all values in between are also present. Examine the 11 diagonals that run upward from left to right. In any diagonal the numbers are all the same. The most common sum is seven in the center diagonal. So we may count the cells for each diagonal to get the predicted frequency of occurrence for the various sums. I list these frequencies.

Sum:	2	3	4	5	6	7	8	9	10	11	12
Frequency:	1	2	3	4	5	6	5	4	3	2	1

The total of the frequencies is 36—there are just 36 cells. So these are the frequencies of the different sums we would predict for 36 throws of two dice.

Since I made six times that number of throws (6 X 36 = 216 throws), we will need to multiply the predicted frequencies by six. Now we can compare prediction and experiment:

Sum:	2	3	4	5	6	7	8	9	10	11	12	Total
Prediction:	6	12	18	24	30	36	30	24	18	12	6	216
Experiment:	9	11	13	34	32	39	25	21	26	12	4	216

The results are also shown graphically below. The broad clear blocks have heights proportional to the experiment values. The solid black blocks have heights corresponding to the theoretical prediction. The predicted values are symmetrical, and adjacent values differ by six. The experimental results seem to follow the pattern set by theory fairly well. I am sure most of you are ready to ask if this is the sort of agreement we can expect between prediction and experiment? Is the disagreement enough to make us question the validity of

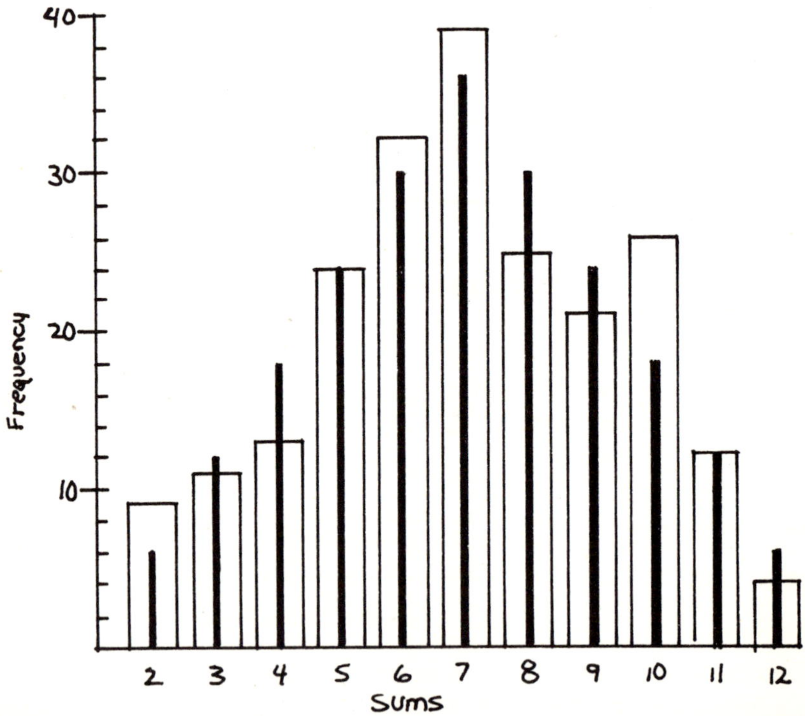

PREDICTED AND EXPERIMENTAL RESULTS FOR SUMS OF TWO DICE

the dice or the technique of making the throws? Mathematics makes it possible to answer that question. And the answer here is that the disagreement is entirely reasonable.

I then looked at the data in another way to check the possibility that some of the dice might be imperfect. The five dice, little white, big white, red, green and orange were rolled together. Three experiments were made, each consisting of 36 rolls giving 108 rolls for each die. The data are shown in the table below.

I made counts of each face for each die, giving the results at the top of p. 64. Since each die was rolled 108 times and has 6 faces, we

RECORD OF 108 THROWS OF FIVE DICE

Set 1	L	153436	546642	524364	113115	121656	112534
	B	636631	213462	622456	114133	531215	244264
	R	223563	624236	235522	231653	651352	531245
	G	441343	451133	146132	142336	654544	522556
	O	533624	231162	326411	332541	144334	361243
Set 2	L	142366	334353	54642	525245	412426	133414
	B	425342	443323	341441	153113	563654	443351
	R	342415	254541	266434	646615	311514	164265
	G	363435	613352	232564	524511	513255	153324
	O	633236	556614	644545	234422	265442	235654
Set 3	L	516653	615346	614541	516166	632465	115236
	B	245614	456342	444511	233616	646141	331231
	R	631444	553643	562244	662121	542625	536326
	G	464533	526561	253423	423454	431414	551254
	O	542656	144153	263112	232636	545335	114234

↑ Face Shown

predict that each face will appear 108/6 = 18 times. We count only three entries of 18, the entries varying from as low as 10 (face 6 on G) to as high as 25 (face 4 on B). It is worth counting up the various entries. There are only three out of the 30 that differ by more than five from the predicted 18. The results do appear to be grouped quite closely and symmetrically around our prediction.

Since there is such a range of observed values (from 10 to 25) and only 30 counts to check against in our table, we again resort to a grouping technique. Suppose we combine the results for the odd-numbered faces (1, 3, 5) and do the same for the even-numbered faces (2, 4, 6). Now make a table of these sums.

TIMES EACH FACE SHOWED IN 108 ROLLS OF 5 DICE

Face	1	2	3	4	5	6
L	22	13	16	18	18	21
B	21	14	22	25	11	15
R	13	21	16	18	20	20
G	15	15	22	23	23	10
O	14	19	22	22	15	16
Sum	85	82	98	106	87	82

Die	L	B	R	G	O	Totals
Odd sums	56	54	49	60	51	270
Even sums	52	54	59	48	57	270
Totals	108	108	108	108	108	540

Let's restate what we are doing. Each die was thrown 108 times, thus presenting us with 108 faces. Faces with one, three or five dots count as odd faces. Those with two, four or six dots count as even faces. Half of the six faces are even, and half are odd. Our best guess is that approximately half of the throws will show an odd face and half an even face. The big white die did exactly this. The most uneven division came from the green die—60 odd and 48 even faces. Even this is only six away from the predicted 54 and 54. The interesting thing is that anything up to ten away would be considered acceptable. Although the formula for this allowable departure is simple, I leave it out because one formula leads to another.

A rather astonishing outcome results from summing the odds and the evens for all five dice. These row totals have come out *exactly* equal. A departure from 270 by as much as 20 would not be unreasonable. There is a rather small chance (about three percent) of getting this exact division into two equal parts. I am glad this happened, because it leads to a very important point. We expect the variation to stay within certain limits (for good dice), but it is also expected that some variation will occur. If the variation is unexpectedly small for any continued string of experiments, suspicion will be aroused immediately in the mind of a statistician. If none of the results for the five dice had departed by more than three from prediction, it would have been a signal for us to worry about the data. Suppose the initials L, B, R, G and O referred to different pupils. I'll be quite frank. If none of the five differed by more than three from the expected 54, I would be strongly inclined to conclude that the pupils had "made up the figures" instead of really doing the experiment.

We can go over again one way the statistician checks up on the data. Take the 30 results in the table on p. 64 where 18 was the prediction. Just as in Chapter 2, the procedure is to take the difference between the number found and the predicted 18. Square the difference and divide by the prediction. Then add up all 30 quotients. Again this sum is called Chi square, and tables show that for this experiment it is quite reasonable for the answer to be anything between 13 and 37. These data gave a Chi square of 25.4, just about midway. When all the dice are well balanced, we can say that for about half of such experiments the variation will exceed what we observe in the data, and for the other half, the variation will be less. To sum it all up, ***there must be variation but not too much and not too little***!

Poisson's formula

7

Years ago the Prussian army (in what is now Germany) kept records of how many cavalry soldiers were killed by horse kicks. Ten regiments kept these records over a 20-year period. Thus there were 200 regiment-years of records. During this period 122 men were killed by horse kicks, an average of 0.61 or less than one man killed in each regiment per year. In one regiment, however, four men died from horse kicks in a single year!

Although we must expect the number of deaths to vary somewhat from year to year, we might wonder if as many as four deaths could occur due to chance. Perhaps the men in that regiment were cruel to their horses or were especially careless. Thanks to a Frenchman named Poisson, we can calculate whether the occurrence of four deaths in one year is extremely unusual. In fact, Poisson's formula is very useful for answering many questions relating to counts of such "rare events."

To use Poisson's formula we must be interested in counting how many times some event occurs. For example, the Prussians counted the number of deaths caused by horse kicks. In a cavalry there must have been numerous opportunities for the horses to kick the soldiers, but luckily for the men, the chance of being fatally kicked was very low. This is precisely the type of situation to which the Poisson formula applies—a situation where there are many opportunities for some event to happen, but the chance of it happening at each opportunity is very small.

To use Poisson's formula all we need to know is the average number of events to expect. For the Prussian cavalry it was 0.61 deaths from horse kicks per regiment per year. Once we know this average rate, we can compute the chance of no events occurring, one event, two events, and so on. Thus we can predict the number of regiments that can be expected to report no deaths, one death, two deaths, or even four deaths. The formula is:

Number of Events	0	1	2	3	4	etc.
Prediction Formula	F	Fm	$\frac{Fm^2}{1 \times 2}$	$\frac{Fm^3}{1 \times 2 \times 3}$	$\frac{Fm^4}{1 \times 2 \times 3 \times 4}$	etc.

We need to know the values for m and F. The average number of events that occur in one time period is m. For the Prussian horse kicks m = 0.61. The value of F is a little more complicated. It is the reciprocal of the quantity 2.718 raised to the power m. The more mathematically advanced reader might recognize $F = e^{-m}$ as the usual way of writing it, where e = 2.718 is the base of the natural logarithms. Now m is usually a decimal, but there are plenty of mathematical tables which tabulate the value of e^{-m} for different values of m. Alternatively, you can compute F by using logarithms. For the horse kicks m = 0.61 and F = .5435. The Poisson formula predictions can be compared with the actual record in the following table.

COMPARISON OF HORSE-KICK DEATHS AND POISSON PREDICTION

No. of men killed per regiment year	0	1	2	3	4	5	Total
Poisson formula	0.5435	0.3315	0.1010	0.0205	0.0030	0.0005	1
Poisson Prediction for 200 regiment-years	108.7	66.3	20.2	4.1	0.6	0.1	200
Observed number of regiment-years	109	65	22	3	1	0	200

The agreement between the observed and expected values is rather spectacular, which is probably the reason this example is used so often as an illustration. The important thing to notice is, that if 0.61 men per year per regiment is considered a good operating record, then we must *expect* as sad a result as four deaths in one of the ten regiments over the period of 20 years. If we had 100 regiments, we would have to expect that in one year some unlucky regiment would lose five men sometime during the 20 years.

There are numerous instances in science, industry and business where the Poisson distribution has been found to be a useful way of examining data. Long ago in 1926, Frances Thorndike, a member of the Bell Telephone Laboratories, published data for 32 examples. About half of these have to do with telephone operations, such as wrong numbers, and cut-offs. The remaining examples are extremely diverse. Among them are counts of alpha particles emitted by a radioactive source, lost articles, fires, and the number of comets observed per year. The distribution is also useful to department and grocery stores as a guide to the number of particular items that should be on the shelves Monday morning so that the chance of disappointing a customer later in the week is held to a chosen small level. I used the distribution once in a study of the number of forgotten passes when men reported for work at a large industrial plant.

Perhaps you can think of a case where the Poisson formula could help you predict how frequently something will occur. Lightning bolts per minute during a storm, the average number of births per month in your county this year, and the number of traffic accidents per week in your community are three examples.

For data that follow the Poisson formula, you can predict within certain limits how many events are likely to occur in some future time period. An especially easy formula arises when m is greater than four or five. Then you can predict that during the next time period, between $m - 2\sqrt{m}$ and $m + 2\sqrt{m}$ events will occur. For example, if on the average there are nine automobile accidents per week in your area, you can predict that there will probably be between three and 15 accidents next week. If the data actually follow the Poisson model, your prediction will be right 19 times out of 20. You can also point out to people that they shouldn't be too surprised when the number of accidents varies from three to 15 accidents per week. You might try to gather some data and see how well the Poisson formula agrees with your data. If it doesn't, perhaps you can figure out why. Sometimes observations don't agree with the Poisson prediction because m is changing. For example, the average number of accidents might be higher during bad weather. Or, sometimes events come in clusters. Deaths from automobile accidents frequently come in clusters since many times several people die in one accident. Of course, the possibility does exist that the disagreement may be due just to chance.

Measuring the thickness of paper

8

Years ago while lecturing to a class of young science students I asked each pupil to measure the thickness of the paper in his chemistry book. The instructions were to open the book at random near the beginning and also at random somewhere past the middle of the book. This stack of pages was held firmly while the thickness of the bundle was measured with a millimeter scale. The page number at the start of the bundle was noted and likewise the page number following the bundle. By custom right-hand pages carry odd numbers, but the difference between the two page numbers which gives the number of pages, will be an even number. The number must be even because when divided by two you get the number of sheets of paper, which has to be a whole number. The number of sheets was divided into the thickness in millimeters of the bundle, and we got the thickness of a single sheet of paper.

I obtained from each pupil the results for four trials. The real reason for detailing this procedure was to make sure that each trial was not influenced by any preceding measurement. If the same numbers of pages are used in each trial, the thickness readings are sure to be influenced by memory, and the same value is apt to be reported for each trial. The results were published in a small book called *Experimentation and Measurement*.*

Recently I ran across a book that describes another technique. A book is opened at random and a bundle about half an inch thick is held between the fingers. The measuring scale is held across the bundle and pages added or removed until the bundle appears to be equal to one-half inch. The page numbers are read as described in the first technique. Dividing 0.50 inches by half the difference in

*W.J. Youden, ***Experimentation and Measurement***, National Science Teachers Association, Washington, D.C. 1962, (127 p.).

the page numbers gives the average thickness of a single sheet. Repeat trials vary depending on how firmly the bundle is held. Of course, you must not peep at the page numbers until you have finished adjusting the bundle to half an inch.

Paper makers follow still a third technique. A fixed number of sheets of paper are stacked on a stand. A rod that applies a standard pressure is brought to bear on the top of the stack, and the thickness of a sheet is read from a dial on the instrument.

Fairly accurate readings can be made by either of the first two techniques—good enough to compare the paper used in different books. You will probably find it necessary to take the average of several readings.

I include these paper experiments because the measurements differ from the type of data taken with the marbles, the coins, the cards, the seeds and the dice we have discussed so far. The data for all those experiments consisted of counts, that is whole numbers. The paper data are different. The stack may be a little less than half an inch. One more sheet may be a little too much. The first technique described shows clearly that the quotients obtained for the successive trials give decimal values. The mathematical models used previously are not suited to our paper measurements. Most measurements in science are like this paper experiment, and every scientist must learn how to interpret such measurements, whether they may be on blood pressure, yield of grain, strength of alloys—the list is endless.

What is fascinating is that one mathematical model serves for nearly all such situations. The mathematics corresponds to a bell-shaped, symmetrical curve. The center point of the curve is the average of the measurements. The width of the curve is related to the scatter of the measurements around the average. The height of the curve is proportional to how frequently measurements have the various values. We must have some measurements available before we can begin to make predictions. We will not go into any details, but the reasoning follows the same way of thought we have found useful in studying data that consist of counts.

Given some preliminary measurements, we are in a position to make an approximate mathematical model. Then we can make predictions about what we might expect to get if we make more measurements on the same paper using the same technique. If measurements made on another source of paper fall outside out pre-

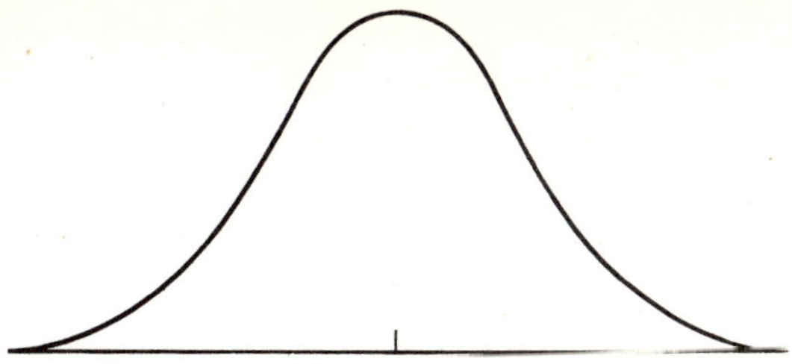

dictions, we are in a position to give up the idea that this new source of paper is identical to the paper whose measurements we used to make our mathematical model. The size of the departure from expectation determines our choice or decision. If the departure exceeds what we would consider acceptable for more measurements made on the first paper, we are justified in claiming that the second paper has a different thickness.

Reluctantly I refrain from giving a detailed set of data. I do urge that you make enough measurements to convince yourself that while measurements do scatter among themselves, they also tend to cluster around their average. Record the number of pages you find each time you make a bundle which appears to be one-half inch thick. These numbers can be divided into 0.50 inches, or used directly. Either way, an approximately bell-shaped curve should appear.

One good way to check your work is to compare it with measurements made by someone else. If the disagreement is as pronounced as we found between the glass and the clay marbles, the question arises as to who is more nearly correct. If a whole group of people supply measurements made on the same book, you may be able to form a judgment by noting where your readings fall in a table made from all available data. Should all your measurements fall in either end of the distribution, it is reasonable to surmise that your technique of making readings is in some way different from the combined group's performance.

There is another method you should consider when checking your performance. Substitute a metric scale for the inch scale. Make your readings for the numbers of pages required to give a stack one centimeter thick. A half-inch is 1.27 times as large as one centimeter. So if you multiply your average number of pages for one centimeter by this factor, you should check the result you got for one half inch.

There is much more that could be said about this subject, but I think the time has come to end our brief journey by taking a quick look back over what we have learned.

What have we learned?

9

There is no really satisfactory way to convey the idea represented by the word ***random***. One reason for emphasizing the importance of getting your own data is that you will continue to build up your own concept of ***random*** on the basis of your own experience. Certainly one may readily admit that successive tosses of a coin will produce heads and tails in a random order. What is this random order? I have often tried the following experiment on friends. First, I ask a friend to copy down a series of at least one hundred digits taken from a published table of random numbers. Then I have him prepare an equally long series of his own concoction (just writing the digits in what he considers to be a random order). I never have any difficulty in picking out the series made up by my friends. Another game consists of asking him to write down a hundred or so T's and H's (for tails and heads) in what he believes to be a random order. Almost always these concocted series are too regular and fail tests for randomness. The mind appears incapable of mimicking the tosses of a coin. The difficulty in specifying what we mean by randomness is sometimes avoided by saying that a random order is one that is generated by a random process!

There is one thing we can say with confidence. Suppose that we do have a very long series of T's and H's in a random order. Then we can say that it is impossible to devise a predictive rule to select out a portion of the series in which the ratio of T's to H's will differ from that in the complete series. Taking every third number won't do it. Taking the next letter after each time you encounter two heads together will not give a subset with a larger proportion of T's than that in the original series. For this reason gamblers at Monte Carlo's roulette wheels never succeed in devising a winning "system" by studying a long record of previous plays.

I hope you have seen in the preceding chapters that the element of randomness is indispensable in planning experiments and in the evaluation of results. I can only reiterate that a little experience is better than a lot of words if you want to understand the emphasis placed on the idea of randomness.

As soon as we decide on an experiment the temptation always comes to rush right into the laboratory. There are good reasons for going slowly at this point. Take time to make the same amount of work produce a more informative experiment. Look for natural groupings or deliberately build them into the program. The ESP program I outlined in Chapter 5 is a good example. Do take time to be sure that you will be able to make some predictions and that you see a way to compare experimental results with your predictions. So many times in my work as a consultant, workers have brought their data to me. They ask, "What do these data mean? What can I do with these data?" And my stock reply is "Tell me ***why*** you did the experiment." This late stage is never a satisfactory time to answer that question. The question should be faced before doing the work.

In the games we have played we were able to make predictions and then compare them with results obtained experimentally. No effort whatever has been made to pick situations where predictions and experiment would conflict. But, I suppose, as I look back over the years, I should not be surprised at what we found. The clay marbles were defective. The bingo cards showed signs of tampering. The penny experiment suggested poor technique. The seed data indicated a shortcoming in the equipment. And, the ESP researchers' card data has failed to convince many scientists. Surely if these simple situations are so vulnerable, we should be most careful in the more intricate researches of modern science. We should build what could be called redundancy into our researches. Just as code messages make provision to straighten out ambiguities that arise from errors in transmission, so should our experiments provide an internal check on the work.

I have often wondered what I would have done if that automatic card selection in my make-believe ESP experiment had shown an excess of face cards impressively greater than predicted. Would I believe that the explanation was to be found in some mystifying manifestation of ESP? Equally strange effects have been claimed. I am sure, however, that I would first verify all my counts and all my arithmetic. If these were in order I would go over the data carefully to see if the excess of face cards was concentrated in some part of the experiment. If the excess was spread over all the work, I would suspect my own shuffling or my technique. I would check to see if the pack contained just twelve face cards and 36 cards with spots. At this point I would probably repeat all the work with a new deck and take even greater care in the work. What I did next would depend on the results obtained in this repeat program.

Suppose that the excess of face cards appeared rather suddenly, say on the third day. Since I did this experiment while convalescing in a hospital, I might wonder if the night nurses had "borrowed" my cards. They might have been returned unshuffled and have pronounced groupings as a result of their game. A discreet inquiry would be in order. If the surplus cards appeared in the morning session after taking medication, I would check with the doctor about the possible effects of the drug. Should the excess face cards be mainly red ones, I think I would get two packs with the same backs. Then I could replace the black face cards with red ones. Should I take out six red spot cards and replace them with black ones to maintain equal numbers of red and black? An interesting question.

The whole experiment would have to be repeated using this modified pack. I am sure I would be startled if this experiment gave an even larger excess of face cards. I would then find someone else willing to do the same experiment with his own cards. Should he, too, find an excess of red cards and his second experiment also checked mine, we would be ready to admit that we were baffled. I imagine that we would prepare a report for publication and include our data. We would ask How could such a system of selecting cards possibly consistently give more red face cards than predicted? We would say that we had no explanation for the mystery and ask others to study the problem.

All of the above discussion of an imagined situation drives home one important point. The built-in plan would make it possible to narrow down the likely sources of any excess of face cards. The plan would guide us in further work, and make it possible for us to write a more satisfactory report.

One of my purposes has been to discuss the *scientific approach* to the search for knowledge about the world we live in. Earlier scientists have learned a lot, often in needlessly tedious ways. Using this hard-bought knowledge we conceive a new inquiry. Much depends on taking time and care in formulating this inquiry. We make a conscious effort to state an appropriate hypothesis. This hypothesis provides us with the means for making predictions. Thus we can compare the experimental results with our predictions. We usually have three choices. The results may be sharply contrary to our prediction, leading us to abandon our hypothesis unless we can detect some fault in our equipment, materials or technique. The second choice is made when the data agree excellently with prediction. We conclude, ***for the time being***, that we do have at least a better understanding of the problem. The third choice comes when the disagreement of data and prediction is not very serious. The usual course then is to accumulate more data to assist in choosing between the first two choices.

I have chosen to discuss simple games rather than actual research data because it is very important that you be able to generate your own data. Many times my first-hand ***experience*** with data has led me to detect a blunder in a formal evaluation of data. Either a formula inappropriate to the situation has been used or an error in arithmetic has been made. The interpretation of real research data usually requires a lot of knowledge about the research problem. Rather expensive equipment and facilities are needed in research. The problems we have discussed, however, need only simple items already available or easily obtained. These simple objects can be used to simulate research. The ESP study with the cards in Chapter 5 is a good example. Making counts is a lot easier than making readings with expensive instruments. The ***logic*** used to examine and to evaluate the data is the same whether the data are counts or very precise measurements. Understanding this logic seems to me to be much more important than presenting a lot of definitions and mathematical formulas.

Appendix

Table A: 70 different orderings of 4 N's and 4 C's

NNNNCCCC

NNNCNCCC

NNNCCNCC

NNNCCCNC

NNNCCCCN

NNCNNCCC

NNCNCNCC

NNCNCCNC

NNCNCCCN

NNCCNNCC

NNCCNCNC

NNCCNCCN

NNCCCNNC

NNCCCNCN

NNCCCCNN

NCNNNCCC

NCNNCNCC

NCNNCCNC

NCNNCCCN

NCNCNNCC

NCNCNCNC

NCNCNCCN

NCNCCNNC

NCNCCNCN

NCNCCCNN

NCCNNNCC

NCCNNCNC

NCCNNCCN

NCCNCNNC

NCCNCNCN

NCCNCCNN

NCCCNNNC

NCCCNNCN

NCCCNCNN

NCCCCNNN

This gives 35. The other 35 can be obtained by replacing all N's by C's and all C's by N's.

Table B: Number of ways of selecting different objects.

Example 1.

If there are four different colors, say red, blue, green and yellow, then there are (4 X 3)/(1 X 2) = 6 different ways to choose two different colors. These are

1. red, blue
2. red, green
3. red, yellow
4. blue, green
5. blue, yellow
6. green, yellow

Example 2.

With five objects (numbered 1, 2, 3, 4, and 5) there will be (5 X 4 X 3)/(1 X 2 X 3) = 10 different ways to choose three different numbers. They are (1, 2, 3); (1, 2, 4); (1, 2, 5); (1, 3, 4); (1, 3, 5); (1, 4, 5); (2, 3, 4); (2, 3, 5); (2, 4, 5); (3, 4, 5).

About the Author

The highlights of W.J. Youden's career are summarized in *The American Statistician* (Vol. 25, June 1971, p. 51), and his contributions to the theory of statistics are discussed in *The Annals of Mathematical Statistics* (Vol. 43, 1972, pp. 1035–1040). The January 1972 issue of the *Journal of Quality Technology* was devoted entirely to his work, reprinting ten of his papers and giving a complete bibliography of his writings and other relevant materials. *Fourteen* of Youden's other papers were reprinted in *Statistical Concepts and Procedures*, National Bureau of Standards Special Publication 300, Volume 1, 1969 (U.S. Government Printing Office, Washington, D.C., 20402), and a collection of bimonthly articles by Youden extending over a six-year period (1954 to 1959) was published under the name *Statistical Design* by *Industrial and Engineering Chemistry*.

One of Youden's publications that may be of particular interest to readers of *Risk, Choice and Prediction* is a 1962 paperback called *Experimentation and Measurement*. This very popular 128-page book was published by the National Science Teachers Association in the *Vistas of Science Series* and is available from the National Education Association of the United States, 1201 Sixteenth Street, N.W., Washington, D.C., 20036.